Mechanisms and Machine Science

Volume 64

This book series establishes a well defined forum for monographs, edited Books, and proceedings on mechanical engineering with particular emphasis on MMS (Mechanism and Machine Science). The final goal is the publication of research that shows the development of mechanical engineering and particularly MMS in all technical aspects, even in very recent assessments. Published works share an approach by which technical details and formulation are discussed, and discuss modern formalisms with the aim to circulate research and technical achievements for use in professional, research, academic, and teaching activities. This technical approach is an essential characteristic of the series. By discussing technical details and formulations in terms of modern formalisms, the possibility is created not only to show technical developments but also to explain achievements for technical teaching and research activity today and for the future. The book series is intended to collect technical views on developments of the broad field of MMS in a unique frame that can be seen in its totality as an Encyclopaedia of MMS but with the additional purpose of archiving and teaching MMS achievements. Therefore the book series will be of use not only for researchers and teachers in Mechanical Engineering but also for professionals and students for their formation and future work. Indexed in SCOPUS, Ei Compendex, EBSCO Discovery Service, OCLC, ProQuest Summon, Google Scholar and SpringerLink.

More information about this series at http://www.springer.com/series/8779

J. C. García-Prada · C. Castejón
Editors

New Trends in Educational Activity in the Field of Mechanism and Machine Theory

2014–2017

Springer

Editors
J. C. García-Prada
Mechanical Engineering Department
Universidad Carlos III de Madrid
Leganés, Spain

C. Castejón
Mechanical Engineering Department
Universidad Carlos III de Madrid
Leganés, Spain

ISSN 2211-0984 ISSN 2211-0992 (electronic)
Mechanisms and Machine Science
ISBN 978-3-030-00107-0 ISBN 978-3-030-00108-7 (eBook)
https://doi.org/10.1007/978-3-030-00108-7

Library of Congress Control Number: 2018953300

This Springer imprint is published by the registered company Springer Nature Switzerland AG
The registered company address is: Gewerbestrasse 11, 6330 Cham, Switzerland

ISEMMS 2017

Introduction

The second International Symposium on the Education in Mechanism and Machine Science (ISEMMS 2017) is a new event on Mechanism and Machine Science (MMS) education. The Executive Council of IFToMM approved and established the ISEMMS symposium to strengthen the research in new educational technologies with a periodicity of four years. The first conference was held in Madrid at the University Carlos III de Madrid (UC3M) during 13–14 June 2013.

One of the main tasks of The Permanent Commission (PC) on Education of IFToMM is to propagate new technologies applied to education in MMS and stabilize a forum where academic people interchange experiences in the field of MMS. This forum is necessary to answer the changing learning and teaching environment in higher MMS education. In this symposium, new teaching and learning methods within MMS had been proposed and presented with applications in the classroom.

The conventional engineering teaching currently is enhanced with new technologies and methods included by and for the new generations. The goal is to obtain graduates with relevant competencies for industry stakeholders or research institutions.

We had selected 26 papers for presentation and publication in this symposium. The programme of ISEMMS 2017 has included technical sessions with oral presentations, two plenary sessions and social activities that are enhanced to share experiences, discussions and let all the participants start new collaborations among them. The contributions cover topics related to MMS in the engineering programme, methodology, applications, research, experiences and new trends.

During the conference, we were honoured by the talk of the vice-president of Strategy and Digital Education of the Universidad Carlos III de Madrid, Prof. Delgado-Kloos, in the plenary session *"Education in Modern Times"*. The plenary offered to the assistants the routes towards the university education in future.

One of the more interesting proposals to develop is the *MMS Teacher's Corner* to be aware of the new teaching tools of mechanical engineering to optimize the time they dedicate to their students. We hope the next symposium will present the results and conclusions of these projects as teamwork with a wide spectrum of universities.

We thank the authors for their excellent presentations and interesting discussions about the contributions. The works presented in this book and the ideas, experiences and developments surely will improve the quality of the education in MMS: applying new engineering education models and improving new engineering graduate subjects.

The editors would like to express their grateful thanks to the members of the steering committee of ISEMMS 2017 who have collaborated for the success of the symposium, especially Professors G. Carbone (Italy), O. Egorova (Russia), P. Flores (Portugal), M. L. Husty (Austria), E. Krylov (Russia), E. Lovasz (Romania), V. Petuya (Spain), P. Fanghella (Italy), and M. Hüsing (Germany).

Also, we thank the Universidad Carlos III of Madrid, IFToMM, AEIM (Spanish Association of Mechanical Engineering), Mechanical Department of UC3M and the Instituto de Investigación Pedro Juan de Lastanosa of UC3M for providing the financial support of ISEMMS 2017.

We thank the members of UC3M-MAQLAB research team for the great effort they put during the preparation and holding of the symposium. They have done a very good job, especially A. Bustos y E. Corral.

Finally, editors would like to remark the invaluable support of the president of IFToMM, Prof. Marco Ceccarelli, with the continuing help to achieve success in ISEMMS 2017.

Juan Carlos García-Prada
Cristina Castejón

Organization

International Scientific Committee

Mario Acevedo, Mexico
Giuseppe Carbone, Italy
Salvador Cardona, Spain
Cristina Castejón, Spain
Marco Ceccarelli, Italy
Olga Egorova, Russia
Pietro Fanghella, Italy
Paulo Flores, Portugal
J. Carlos García-Prada, Spain
Antoni Gronowicz, Poland
Manfred L. Husty, Austria
Dae-Eun Kim, Korea
Eduard Krylov, Russia
Erwin Lovasz, Romania
Juana Mayo, Spain
Karol Miller, Australia
Jose Luis Ocaña, Spain
Alba Perez-Gracia, EEUU
Victor Petuya, Spain
Riccardo Rubini, Italy
Yukio Takeda, Japan
María Tomas-Rodriguez, UK

Organizing Committee

Juan Carlos García Prada (Chair)	Universidad Carlos III de Madrid
Cristina Castejón (Vice Chair)	Universidad Carlos III de Madrid
Jesús Meneses (Organizing)	Universidad Carlos III de Madrid
Higino Rubio (Organizing)	Universidad Carlos III de Madrid
Mª Jesús Gomez (Organizing)	Universidad Carlos III de Madrid
Eduardo Corral (Secretary)	Universidad Carlos III de Madrid
Alejandro Bustos (Secretary)	Universidad Carlos III de Madrid

Contents

Mechanical Engineering Dducation: Experiences

MMS in the Engineer Program: Methodology

Method of Closed Vector Contours for Teaching/Learning MMS

B. Kosenok[1(✉)], V. Balyakin[1], and E. Krylov[2]

[1] Samara State Aerospace University, Samara, Russia
borkos@yandex.ru
[2] Kalashnikov Izhevsk State Technical University, Izhevsk, Russia
649526@mail.ru

Abstract. The paper discusses the basics of the mathematical modeling method and makes the substantiation of the necessity of the method development for solving more complex mechanisms with the generation of vector models with the use of "intermediate" arguments and modules with function links. The main advantages of this approach are shown for the study of the theory of mechanisms and ma-chines by students.

Keywords: Vector · Contour · Model · Modules · Invariance

The method of closed vector contours [1, 2], is successfully used to make a mechanism models in the analysis and synthesis problems, and also to the mechanism design problems (synthesis of the structural or kinematic scheme of a mechanism that provides the required transfer function).

The basis of the method is vector closed contours consisting of separate vectors. Vector is a directed segment with parameters r_i and α_i (Fig. 1). It is known, that any mechanism structural scheme can be considered as a vector contour (Fig. 2).

The closed vector contours of that kind can be reduced to the elementary ones (modules) having well known and detailed solutions, which allows one to get a system of vector contours using a simple method of selecting the required modules.

The basic vector model constructed on the basis of modules allows one to obtain a solution of certain functions, with model arguments known for each moment of time and the relationships between individual vectors and contours.

The closure condition for the planar m-vector contour:

$$\sum_{i=1}^{m} r_i \cos\alpha_i = 0; \quad \sum_{i=1}^{m} r_i \sin\alpha_i = 0. \tag{1}$$

The closure condition for the spatial m-vector contour is:

$$\sum_{i=1}^{m} r_i \cos\alpha_i \cos\beta_i = 0; \quad \sum_{i=1}^{m} r_i \sin\alpha_i = 0; \quad \sum_{i=1}^{m} r_i \cos\alpha_i \sin\beta_i = 0. \tag{2}$$

J. C. García-Prada and C. Castejón (Eds.): *New Trends in Educational Activity in the Field of Mechanism and Machine Theory*, MMS 64, pp. 3–10, 2019.
https://doi.org/10.1007/978-3-030-00108-7_1

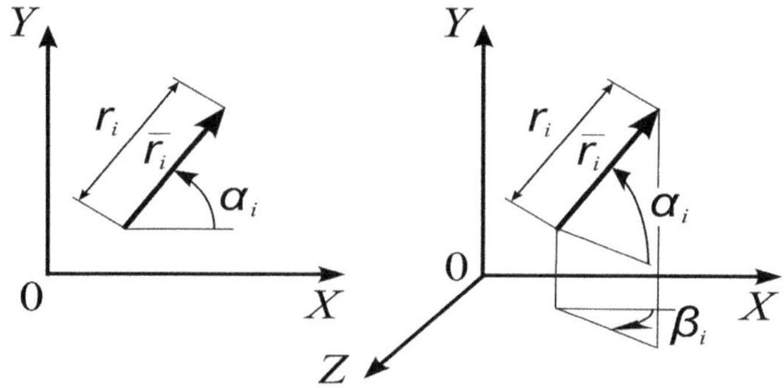

Fig. 1. Parameters of 2-dimensional and 3-dimensional vectors

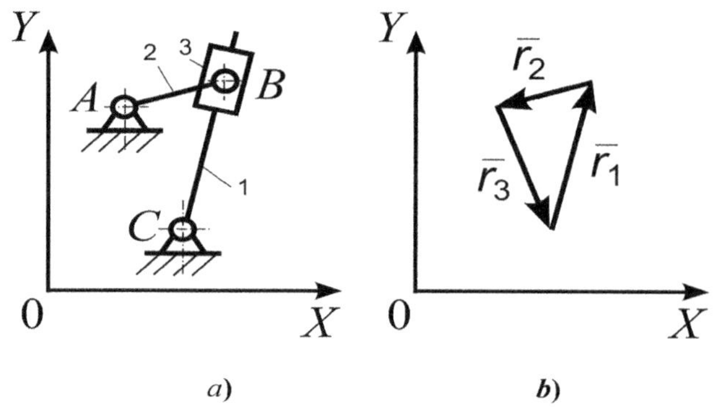

Fig. 2. Structural scheme (a) and vector model (b) of a mechanism

The basic combinations of functions of vector contours, based on the closure condition (1), define four planar elementary modules (PL1–PL4), and twenty spatial (PR1–PR20) from condition (2) [1, 2]. Also, it should be noted that the modules can have several solutions, and the solution of vector modules with function links adds elementary modules with solution options. All eight modules for a plane are listed in Table 1.

Spatial modules are listed in Table 2. The solutions are found for the functions for planar modules PL1–PL4, PL1 s–PL4 s and spatial PR1–PR5, PR7–PR13, PR18, PR19.

The use of vector models in modeling problems of kinematics and dynamics of mechanisms, firstly, allows one to perform a study of the operation of a mechanism flexibly and quickly at the stage of preliminary design, and to make the optimization of the main parameters of the mechanism (link lengths, pressure angles, reduced loads and mass characteristics, reactions in hinges, etc.). Besides, this technique solves the related problems of designing, and makes the mathematical tools predictable and ready to be implemented in software package for the calculation of a mechanism kinematics and dynamics (*KDAM*).

Table 1. Planar elementary vector modules

Module	Functions		Number of solutions	Basic arguments - any vector parameters not equal to the functions of the module							
	u_1	u_2		r_i	α_i	r_j	α_j	r_{il}	α_{i1}	r_{i2}	α_{i2}
PL1	r_i	α_i	1			x	x	x	x	x	x
PL2	r_i	r_j	1		x		x	x	x	x	x
PL3	r_i	α_j	2		x	x		x	x	x	x
PL4	α_i	α_j	2	x		x		x	x	x	x
Vector modules with function link											
PL1 s	r_i	α_i	2			x	x	x	x	x	x
PL2 s	r_i	r_j	2		x		x	x	x	x	x
PL3 s	r_i	α_j	4		x	x		x	x	x	x
PL4 s	α_i	α_j	4	x		x		x	x	x	x

Table 2. Spatial elementary vector modules

Module	Functions			Number of solutions	Basic arguments - any vector parameters not equal to the functions of the module														
	u_1	u_2	u_3		r_i	α_i	β_i	r_j	α_j	β_j	r_k	α_k	β_k	r_l	α_l	β_l	r_m	α_m	β_m
PR1	r_i	α_i	β_i	1										x	x	x	x	x	x
PR2	r_i	α_i	r_j	1			x		x	x				x	x	x	x	x	x
PR3	r_i	α_i	α_j	2			x	x		x				x	x	x	x	x	x
PR4	r_i	α_i	β_j	2			x	x	x					x	x	x	x	x	x
PR5	r_i	β_i	r_j	2		x			x	x				x	x	x	x	x	x
PR6	r_i	β_i	α_j	4		x		x		x				x	x	x	x	x	x
PR7	r_i	β_i	β_j	2		x		x	x					x	x	x	x	x	x
PR8	α_i	β_i	r_j	2	x				x	x				x	x	x	x	x	x
PR9	α_i	β_i	α_j	2	x			x		x				x	x	x	x	x	x
PR10	α_i	β_i	β_j	2	x			x	x					x	x	x	x	x	x
PR11	r_i	r_j	r_k	1		x	x		x	x		x	x	x	x	x	x	x	x
PR12	r_i	r_j	α_k	2		x	x		x	x	x		x	x	x	x	x	x	x
PR13	r_i	r_j	β_k	2		x	x		x	x	x	x		x	x	x	x	x	x
PR14	r_i	α_j	α_k	4		x	x	x		x	x		x	x	x	x	x	x	x
PR15	r_i	α_j	β_k	4		x	x	x		x	x	x		x	x	x	x	x	x
PR16	α_i	α_j	α_k	8	x		x	x		x	x		x	x	x	x	x	x	x
PR17	α_i	α_j	β_k	8	x		x	x		x	x	x		x	x	x	x	x	x
PR18	β_i	β_j	r_k	2	x	x		x	x			x	x	x	x	x	x	x	x
PR19	β_i	β_j	α_k	4	x	x		x	x	x	x		x	x	x	x	x	x	x
PR20	β_i	β_j	β_k	∞	x	x		x	x		x	x		x	x	x	x	x	x

Similar reasons make it possible and necessary to apply the theory of vector modeling in the teaching/learning process, for the design project, and labs on the theory of mechanisms and machines (TMM). If the traditional TMM course [3–5] provides an understanding of the main elements of mechanisms, a mechanism structure, calculation of the kinematical and dynamical parameters of mechanisms, then the use of *KDAM* allows students to understand the fundamentals of analysis and synthesis of mechanisms.

Figures 3, 4, 5 and 6 illustrates the use of *KDAM* during the execution of the *TMM design project.*

1. For the kinematical analysis of the mechanism (Fig. 3).

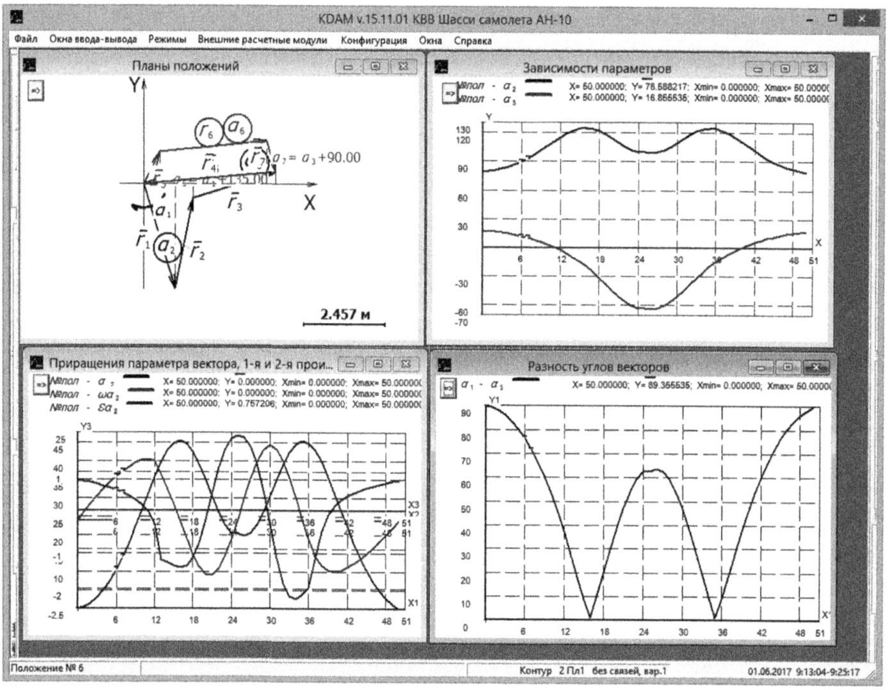

Fig. 3. Example of use *KDAM* for the vector modeling of the mechanism of the landing gear

2. For the dynamical analysis of the mechanism (Fig. 4).

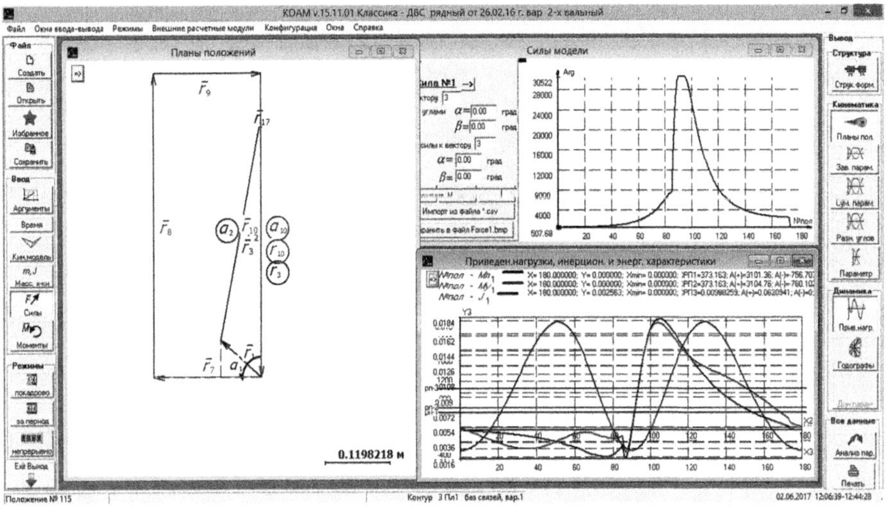

Fig. 4. Example of the KDAM use for finding the reduced loads of the ICE mechanism

3. For the design of the gear train (Fig. 5).

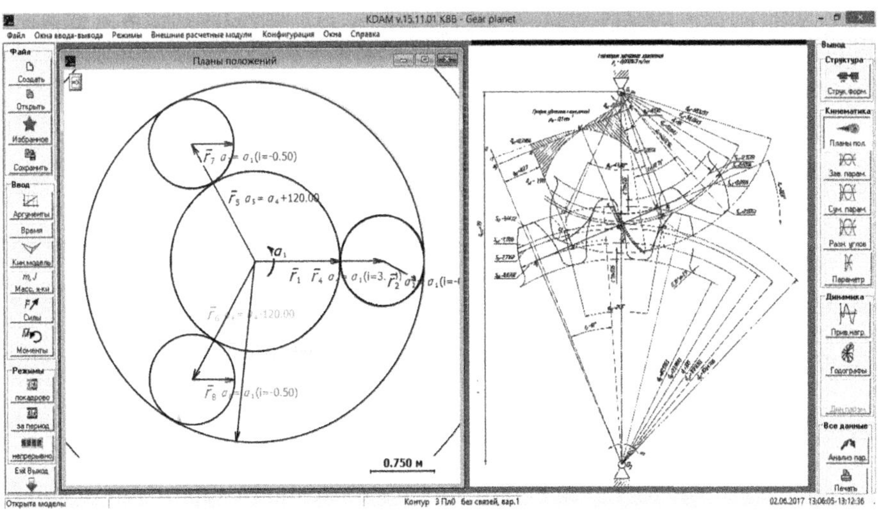

Fig. 5. Simulation of the planetary gearbox in *KDAM*

4. For the design of the cam profile (Fig. 6).

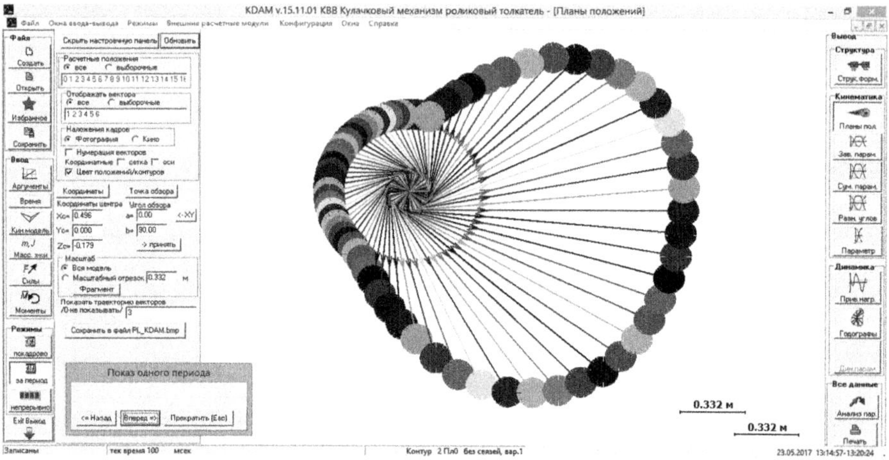

Fig. 6. *KDAM* vector modeling of a cam profile in the mechanism with a roller follower

The course "Kinematics and Dynamics of Internal-Combustion Engine (ICE)" as a highly specialized course allows reach more significant understanding of the problems and challenges facing engine design and manufacturing, and at the same time, in comparison of various schemes of a mechanism, enable students themselves to explore and find it positive and negative aspects (Figs. 7 and 8).

Fig. 7. Engine MAZDA SKYACTIV

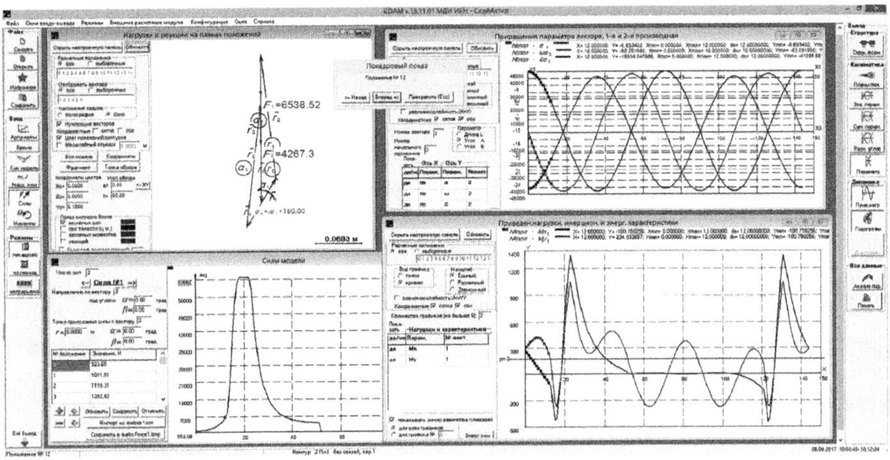

Fig. 8. *KDAM* modelling of the mechanism of Engine MAZDA SKYACTIV

The use of *KDAM* during the execution of the design project "Kinematics and Dynamics of Internal Combustion Engine" [2] is illustrated below.

1. Overview of the given type of engine (gearbox, drive):
   a. Unit appearance, structural scheme, and vector model for kinematics.
   b. Production program and statistics.
   c. Areas of the unit use.
   d. Analysis of weight, economic, and environmental characteristics.
   e. Table of main characteristics.
   f. Possible ways of development.
2. Features of new design in comparison with the classical one:
   a. The comparison of a basic mechanism (working surfaces, heat transfer, sealing's, the number of moving links, mobility, redundant constraints and other).
   b. The list of auxiliary systems, and how they differ from classical ones.
   c. Pros and cons of the systems under consideration.
3. Analysis of kinematical and dynamical perfection of the design considered [6–8] (using KDAM) in comparison with the classical one:
   a. Structural scheme and vector model describing kinematics and dynamics.
   b. If equal volumes of working chambers.
   c. If identical dimensions.
   d. Variants of working cycles (for example, two-stroke, three-stroke cycles).
4. Analysis of strength and durability characteristics.
5. Conclusions.
6. Reference.

For full mastering by students of *KDAM* the set of laboratory works is used, where students consider various methods of kinematical and dynamical analysis, and synthesis of mechanisms.

1 Conclusion

The main advantages of the proposed approach to mastering the theory of mechanisms and machines is that when using *KDAM* software, there is no need for long learning both the vector modeling method itself, and the software product. This is due to the fact that all elements are of the similar type and the learning both aspects are reduced to one academic lesson. The rest of the time is given both to classical methods of calculation and to the use of *KDAM*'s capabilities for a full kinematic and dynamic analysis of the designed mechanisms, which brings the studying closer to real design problems and gives students a more complete idea of the mechanisms, and also the ability to analyze many parameters of the mechanisms, make kinematic and dynamic optimization with minimal efforts with reliable results obtained, which can be useful for them in daily work as a designer or a technical expert.

References

1. Semenov, B.P.: Analytics of Elementary Vector Modules. Moscow Aviation Institute Publishing House, 40 p. (1989)
2. Semenov, B.P., Kosenok, B.B.: Methods and Means of Dynamical Synthesis of Mechanisms of Aviation Power Plant Units, p. 281. Samara Science Center of RAS, Samara (2010)
3. Artobolevsky, I.I.: Theory of Mechanisms and Machines, 4th edn. Science, 639 p. (2009)
4. Savinov, A.P., Yastrebov, V.M., Korobova, N.P.: Theory of Mechanisms and Machines in Aircraft Industry, 173 p. SSAU, Samara (2006)
5. Timofeev, G.A. (ed.): Theory of Mechanisms and machines. Course design, 2nd edn., 169 p. Moscow Bauman State University, Moskva (2012)
6. Kosenok, B.B., Balyakin, V.B.: Study of the dynamic characteristics of a two-cylinder internal combustion engine using vector models. Procedia Eng. **106**, 183–191 (2015)
7. Kosenok, B.B.: Crank-rod mechanism for an internal combustion engine. Russ. Eng. Res. **37** (1), 19–22 (2017). © Allerton Press, Inc., 2017. Original Russian Text © B.B. Kosenok, B.V. Balyakin, I.N. Zhil'tsov, 2016, published in Vestnik Mashinostroeniya, 2016, No. 10, pp. 47–50
8. Kosenok, B.B., Balyakin, V.B., Giltsov, I.N.: The use of vector models to study the dynamic characteristics of an advanced two-shaft internal combustion engine. Procedia Eng. **176**, 37–42 (2017)

Design of Cam Mechanisms with Swinging Roller Follower: The Modern Algorithm-Based Approach

Lyudmilla Chernaya(✉), Andrei Vukolov, and Il'ya Safronoff

Bauman Moscow State Technical University, Moscow, Russian Federation
chernayala@yandex.ru, twdragon@bmstu.ru

Abstract. Modern mechanical engineering and design of the mechanisms and machines are closely connected with development of numerical methods and algorithms. Education is the useful target to apply these methods. Development of the analytical and algorithmic competences is possible only using high technologies through whole course. The modern numerical method on the metric synthesis of cam mechanism is demonstrated in this paper using new algorithmical approach and `MathCAD` realization.

Keywords: `MathCAD` · Cam mechanism · Swinging follower
Numerical methods · Numerical modeling · Roller follower

1 Introduction

In Bauman Moscow State Technical University the generalized methods of design and exploration of machines are expounded within Theory of Mechanisms and Machines (TMM) course [2,12]. In the past the approximate graphical methods [19,20] were mostly used. Now the available calculation resources made it possible to use almost any method [3], including numerical and analytical [16].

The followers of cam mechanisms move under strictly determined trajectory [11] and motion laws [4]. They require precise calculation of the profile's coordinates for successful manufacturing [17,18]. The modern calculation software products like `MathCAD` [14,15] provide such capabilities. They are easy to use, so development and realization of new approaches and algorithms are good challenges for engineering students [7,10]. Here and below the typical cam mechanism with swinging follower and four-phased (rise, return and two dwellings) cycle will being considered as an example.

J. C. García-Prada and C. Castejón (Eds.): *New Trends in Educational Activity in the Field of Mechanism and Machine Theory*, MMS 64, pp. 11–19, 2019.
https://doi.org/10.1007/978-3-030-00108-7_2

2 Target Setting

The mechanism (Fig. 1a) is constrained with following parameters:

- Maximal displacement h_B of the follower's contact point B which determines rising angle β;
- Length l_2 of the rocker 2;
- Phase angles: rise angle ϕ_{1y}, return angle ϕ_{1c} and dwelling angle $\phi_{1\partial}$;
- Motion law of the follower defined graphically;
- Upper limit of the pressure angle $[\vartheta]$;
- Rotation direction defined with sign of the ω_1 value.

The whole task of designing of the cam mechanism can be splitted into three stages [8]:

1. Building of kinematical diagrams and calculating of geometrical characteristics;
2. Metrical synthesis constrained by $[\vartheta]$;
3. Kinematical synthesis—building of theoretical cam profile and envelope surface.

The whole motion law of the follower for the considered mechanism is presented on Fig. 1b. The motion law here defined with coefficients a_1, a_2, a_3, a_4, scales μ and ratio of their values.

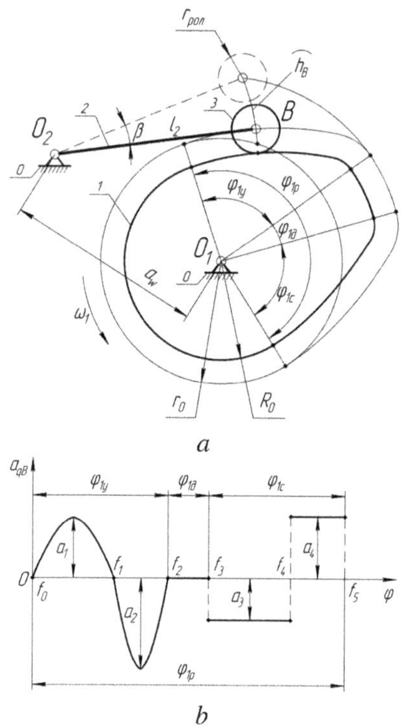

Fig. 1. Cam mechanism and follower's motion law

3 Kinematical Diagrams

According to Fig. 1b the acceleration analog function a_{qB} has discontinuity points of type I at f_3, f_4 and f_5. To satisfy requirements of algorithmical approach the Haeviside function $\Phi(x)$ [9] used to determine discontinuities. Now the following expressions can be written for the followers's motion law on the rising phase:

$$a_{qBy}(\varphi) = \begin{cases} \frac{\varphi}{\varphi_1} \sin\left(\pi, \frac{\varphi}{f_1}\right), \text{ where } 0 \leq \varphi < f_1 \\ \frac{-\pi}{f_2 - f_1} \sin\left[\pi \frac{f_2 - \varphi}{f_2 - f_1}\right], \text{ where } f_1 \leq \varphi \leq f_2 \\ 0, \text{ where } f_2 < \varphi \leq f_3 \end{cases} \quad (1)$$

... and on returning phase where $f_3 < \varphi \leq f_5$ (Fig. 1*b*):

$$a_{qBc}(\varphi) = [-a_3 \cdot \Phi(\varphi - f_3) + (a_3 + a_4) \cdot \Phi(\varphi - f_4) - a_4 \cdot \Phi(\varphi - f_5)] \quad (2)$$

Now the kinematical characteristics: speed analog function $v_{qB}(\varphi)$ and follower displacement function $S_B(\varphi)$ could be obtained.

4 Metric Synthesis

The main constraint of the metric synthesis task is the pressure angle limitation $[\vartheta]$. The formalities for algorithmical approach are the following:

1. All schemes considered as they are built in right Cartesian coordinates S_{xOy} with origin placed into the fixed point O_1 of the cam (Fig. 1*a*).
2. Rotation angle φ of the cam directed conterclockwise is generalized coordinate;
3. Angular velocity analog function ω_{q1} of the cam determines rotation direction (1 for counterclockwise direction).

Let the cam 1 (Fig. 1*a*) rotate and the follower 2 moves by arbitrary trajectory and vector $\overline{v_B} = \overline{v_2}$ of the contact point B's absolute speed is known. Let the Y axis of the S_{xOy} coordinate system with origin in O_1 is parallel to $\overline{v_2}$. Now the pressure angle can be defined as the following function:

$$\tan \vartheta(\varphi) = \frac{v_{qB}(\varphi) - x_B(\varphi)}{y_B(\varphi)} \quad (3)$$

Now we reinterpret (3) for case of swinging follower:

$$\tan \vartheta(\varphi) = \frac{\omega_{q1} \cdot v_{qB}(\varphi) - l_2 + a_w \cos(\varphi_{20} + \varphi_2(\varphi))}{a_w \sin(\varphi_{20} + \varphi_2(\varphi))} \quad (4)$$

where

$$\cos \varphi_{20} = \frac{l_2^2 + a_w^2 - r_0^2}{2 l_0 \cdot a_w}$$

and φ_{20} is initial angle [5] between l_2 and O_1O_2 line at the lower dwelling phase (Fig. 1*a*).

With known dependence between ϑ and φ the phase diagram $S_B[v_{qB}(\varphi)]$ can be built [17] in right Cartesian coordinate system S_{xAy} with origin in A point which coincides with center O_2 of the rocker (Fig. 1). On the diagram (Fig. 2) the O_1 and O_2 points demonstrate possible positions of the cam center [5]. The metric synthesis of the cam mechanism includes obtaining values for a_w and r_0 parameters with known l_2 and $[\vartheta]$. With known direction of rotation of the cam the upper limit of the pressure angle [6,12] can be determined on rising phase $[\vartheta_y]$, returning phase $[\vartheta_c]$ or on both phases for reversible mechanism. Now the lines defined by $[\vartheta]$ have to be determined in S_{xAy}. If coordinates of i-point of the diagram are X_i, Y_i the line passing this point is defined as

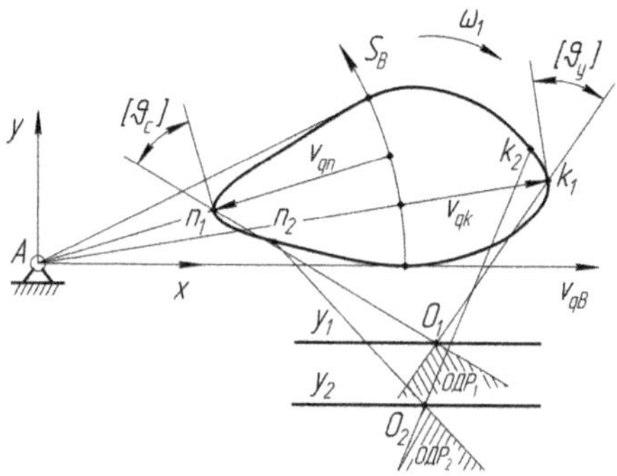

Fig. 2. Phase diagram

$Y_i = k_i \cdot X_i + b_i$ equation. The extreme position of such line is determined on the rising phase with angle $\varphi_2(\varphi) + \pi/2 - [\vartheta_y]$ and on returning phase with angle $\varphi_2(\varphi) + \pi/2 + [\vartheta_c]$. Now:

$$\begin{aligned} k_{yi} &= \tan\left(\varphi_2(\varphi) + \pi/2 + \omega_{q1}\,[\vartheta]\right) \\ k_{ci} &= \tan\left(\varphi_2(\varphi) + \pi/2 - \omega_{q1}\,[\vartheta]\right) \end{aligned} \tag{5}$$

The intersection point O_i between lines on Fig. 2 is defined by equation $k_y \cdot X + b_y = k_c \cdot Y + b_c$. After substitutions we have:

$$\begin{aligned} X &= -\frac{b_c - b_y}{k_c - k_y} \\ Y &= \frac{k_c b_y - k_y b_c}{k_c - k_y} \end{aligned} \tag{6}$$

Now the point O_1 with coordinates X_1, Y_1 is placed where the lines defined by (6), maximal and minimal values of $v_{qB}(\varphi)$ intersect. For each point of the diagram the line could be drawn through O_1:

$$X_i = \frac{Y_1 - b_i}{k_i} - X_1 \tag{7}$$

Now the function $f(x) = X_i(\varphi)$ can be obtained and the lines which constrain the zone of possible location for the cam center are placed above Ax axis for rising phase and below it for returning. The algorithm of cam center point calculation is now determined within the three stages:

- Extremum analysis of the function $X_i(\varphi)$ for coordinates φ_y and φ_c of the most distant lines (Fig. 2);
- Substitution of φ_y and φ_c into (6) for X_2 and Y_2. Now the levels y_1 and y_2 can be placed;
- Obtaining of the pressure angle function $\vartheta(\varphi)$ from (4).

Realization of the considered approach were performed by I. Safronoff using `MathCAD`. Here the fragment of the source code[1] is presented:

$$k_{Vi}(\phi) := \begin{cases} \tan\left(\varphi_2(\phi) + \frac{\pi}{2} + \theta d \cdot \omega_{qk}\right) & \text{if } f0 \le \phi \le f2 \\ \tan\left(\varphi_2(\phi) + \frac{\pi}{2} - \theta d \cdot \omega_{qk}\right) & \text{if } f3 \le \phi \le f5 \\ 0 & \text{otherwise} \end{cases}$$

$$b_{.Vi}(\phi) := Ny(\phi) - k_{.Vi}(\phi) \cdot Nx(\phi)$$

Initial approximations: left extremum

$$\phi_{Vmax} := 100 \cdot deg$$

Given $\quad X_{Vi}(\phi_{Vmax}) = 0.1 \qquad 60 \cdot deg < \phi_{Vmax} < 130 \cdot deg$

$$\varphi_{Vmax} := \text{Minerr}(\phi_{Vmax}) = 100.8 \cdot deg \qquad X_{Vi}(\varphi_{Vmax}) = 3.688 \times 10^{-4}$$

Right extremum:

$$\phi_{Vmin} := 275 \cdot deg$$

Given $\quad X_{Vi}(\phi_{Vmin}) = -0.1 \qquad 250 \cdot deg < \phi_{Vmin} < 300 \cdot deg$

$$\varphi_{Vmin} := \text{Minerr}(\phi_{Vmin}) = 281.933 \cdot deg \qquad X_{Vi}(\varphi_{Vmin}) = -9.843 \times 10^{-3}$$

Intersection point:

$$X_2 := -\left[\frac{b_{Vi}(\varphi_{Vmax}) - b_{Vi}(\varphi_{Vmin})}{(k_{Vi}(\varphi_{Vmax}) - k_{Vi}(\varphi_{Vmin}))}\right] \qquad Y_2 := \frac{k_{Vi}(\varphi_{Vmax}) b_{Vi}(\varphi_{Vmin}) - k_{Vi}(\varphi_{Vmin}) b_{Vi}(\varphi_{Vmax})}{k_{Vi}(\varphi_{Vmax}) - k_{Vi}(\varphi_{Vmin})}$$

a_w and R_{min}:

$$a_w := \sqrt{X_2^2 + Y_2^2} = 0.395 \qquad R_{min} := \sqrt{(Xk - l_t)^2 + Yk^2}$$

Phase diagram obtained from `MathCAD` is presented on Fig. 3*a*. The coordinate system described in Sect. 4 is presented on Fig. 3*b* as $x_K O y_K$ axis. The coordinate

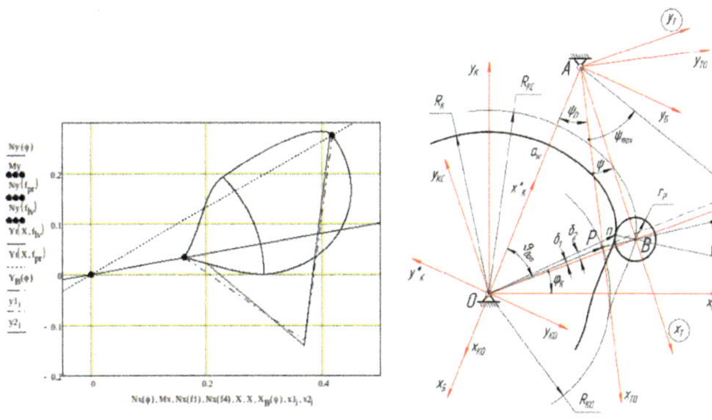

a *b*

Fig. 3. Phase diagram obtained using MathCAD and cam profile with coordinate systems applied

[1] Here and below all `MathCAD` source code listings are written by L. Chernaya and I. Safronoff.

system $x_T A y_T$ is connected to the follower. Angle α_∂ represents pressure angle ϑ from Fig. 2. From phase diagram the minimal cam radius R_{min} now can be obtained.

5 Kinematical Synthesis

Matrix equation of coordinate transform between cam and follower (Fig. 3) is presented below. The transition performs over coordinate systems $S_{x_T A y_T}$ and $S_{x_{KC} A y_{KC}}$:

$$[A_{KcT}(\varphi)] = [A_{KcK}] \cdot [A_{KK^*}(\varphi)] \cdot [A_{K^*K_0}] \cdot [A_{K_0B}]\,[A_{BT}(\varphi)] \tag{8}$$

The final equation of theoretical cam profile with argument φ looks as:

$$K_{cp}(\varphi) = [A_{KcT}(\varphi)] \cdot \begin{pmatrix} l_2 \\ 0 \\ 1 \end{pmatrix} \tag{9}$$

where l_2 is rocker length (Fig. 1a). `MathCAD` realization of matrix calculations described here is obtained by the following code for initial circle R_0:

$X_{k.0}(\varphi) := R_{min} \cdot \cos(\varphi + \omega_{qk} \cdot \varphi_k)$ $\quad$ $Y_{k.0}(\varphi) := R_{min} \cdot \sin(\varphi + \omega_{qk} \cdot \varphi_k)$

theoretical cam profile:

$\psi_\Sigma(\varphi) := \psi 0 + \phi_2(\varphi)$ $\qquad$ $\varphi_{dop} := \operatorname{acos}\left(\dfrac{a_w^2 + R_{min}^2 - l_t^2}{2 \cdot a_w \cdot R_{min}}\right) = 42.854\,\text{deg}$

$$A_{BT}(\varphi) := \begin{pmatrix} \cos(\psi 0 + \phi_2(\varphi)) & -\sin(\psi 0 + \phi_2(\varphi)) & 0 \\ \sin(\psi 0 + \phi_2(\varphi)) & \cos(\psi 0 + \phi_2(\varphi)) & 0 \\ 0 & 0 & 1 \end{pmatrix} \quad A_{KxKo} := \begin{pmatrix} 1 & 0 & a_w \\ 0 & 1 & 0 \\ 0 & 0 & 1 \end{pmatrix} \quad A_{KoB} := \begin{pmatrix} -1 & 0 & 0 \\ 0 & -1 & 0 \\ 0 & 0 & 1 \end{pmatrix}$$

$$A_{KKx}(\varphi) := \begin{pmatrix} \cos(\varphi + \varphi_{dop}) & -\sin(\varphi + \varphi_{dop}) & 0 \\ \sin(\varphi + \varphi_{dop}) & \cos(\varphi + \varphi_{dop}) & 0 \\ 0 & 0 & 1 \end{pmatrix} \quad A_{KcK} := \begin{pmatrix} \cos(\varphi_k) & -\omega_{qk} \cdot \sin(\varphi_k) & 0 \\ \omega_{qk} \cdot \sin(\varphi_k) & \cos(\varphi_k) & 0 \\ 0 & 0 & 1 \end{pmatrix}$$

$TR_C(\varphi) := A_{KcK} \cdot A_{KKx}(\varphi) \cdot A_{KxKo} \cdot A_{KoB} \cdot A_{BT}(\varphi)$ $\qquad$ $r_{rol} := 0.3 \cdot R_{min} = 0.047$

$REZ_C(\varphi) := TR_C(\varphi) \cdot \begin{pmatrix} l_t \\ 0 \\ 1 \end{pmatrix}$ $\qquad$ $X_{kCM}(\varphi) := REZ_C(\varphi)_0$ $\qquad$ $Y_{kCM}(\varphi) := REZ_C(\varphi)_1$

...and for the envelope curve:

$$TURN_v(\varphi) := \begin{pmatrix} \cos(\theta_{dav}(\varphi)) & -\sin(\theta_{dav}(\varphi)) & 0 \\ \sin(\theta_{dav}(\varphi)) & \cos(\theta_{dav}(\varphi)) & 0 \\ 0 & 0 & 1 \end{pmatrix} \quad SHIFT_L := \begin{pmatrix} 1 & 0 & l_t \\ 0 & 1 & 0 \\ 0 & 0 & 1 \end{pmatrix} \quad TR(\varphi) := SHIFT_L \cdot TURN_v(\varphi)$$

The final cam profiles calculated and drawn in `MathCAD` are presented on Fig. 4. The dotted lines represent initial circle R_{K0} and theoretical profile from Fig. 3. The solid ones represent envelope curve (manufacturing-ready profile) and minimal radius R_K. The roller radius r_p is fixed [11]. The algorithm developed by

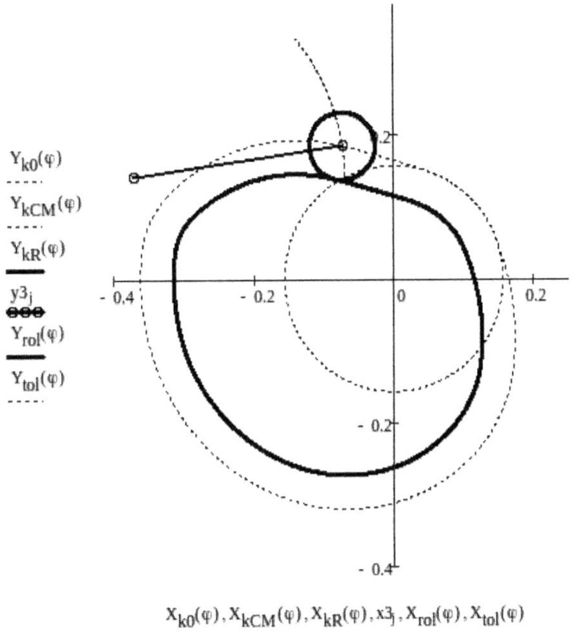

Fig. 4. Cam profiles drawn by MathCAD

I. Safronoff allows to easily connect the numerical methods which `MathCAD` implements to both highly-constrained calculation task of the metric synthesis and geometrical task of profiling.

6 Conclusion

The algorithm-based approach demonstrated above allows the student to use his skills of software development and toolchain building. In Bauman University it links traditional design with new techniques. Theoretical knowledge of the theory of mechanisms and machines provided with lectures could be successfully applied by student using continuous integration with modern software and technologies [1]. The case of `MathCAD` usage quoted in this paper allows to export data into simple format supported by manufacturing solutions using available built-in functions. In the modern system of engineering education the scripts, datasheets and programs developed by students themselves is the way to provide possibility to obtain industrial experience within training. Also the networks and Web-based programming techniques [13] allow university to build open educational space with free distributed workflow.

References

1. Angeles, J., López-Cajún, C.S.: The Computer-Aided Drafting and Manufacture of Cams, pp. 208–227. Springer, Netherlands (1991). https://doi.org/10.1007/978-94-011-3572-6_9
2. Artobolevsky, I.I.: Theory of Mechanisms and Machines. The University Handbook, 4, Remastered edn. Nauka. The Main Editorial of Physical and Mathematical Literature, Moscow (1988). (in Russian)
3. Aziz, R.: Development of an integrated system for cam design and manufacture with graphical user interface. In: CD-ROM Proceedings of the 1996 ASME Design Engineering Technical Conferences and Computers in Engineering Conference, Irvine, CA (1996)
4. Babichev, D., Lagutin, S., Barmina, N.: Russian school of the theory and geometry of gearing: its origin and golden period (1935-1975). Front. Mech. Eng. **11**(1), 44–59 (2016). https://doi.org/10.1007/s11465-015-0360-z
5. Belyaev, A.N.: Analysis and synthesis of cam mechanisms. Handbook for engineering students. Voronezh State Agrarian University, Voronezh (2004). (in Russian)
6. Belyaev, A.N., Klimov, G.D., Sheredekin, V.V.: Design of cam mechanisms. Handbook for engineering students. Voronezh State Agrarian University, Voronezh (2008). (in Russian)
7. Chernaya, L.A.: Kinematical and kinetostatical studies of the flat linkages using Mathcad and AutoCAD systems. Reference manual on the theory of mechanisms and machines. Bauman Moscow State Technical University. Publishing House, Moscow (2017). (in Russian)
8. Chernaya, L.A., Timofeev, G.A.: Theory of Mechanisms and Machines. Handbook on the Course Projects (in Russian) (2017). preprint edn. Bauman Moscow State Technical University. Publishing House, Moscow (2018)
9. Davies, B.: The Laplace transform, pp. 27–38. Springer, New York (2002). https://doi.org/10.1007/978-1-4684-9283-5_2
10. Duma, V.F.: Teaching mechanisms: from classical to hands-on-experiments and research-oriented, pp. 493–501. Springer, Netherlands (2010). https://doi.org/10.1007/978-90-481-9689-0_57
11. Koloc, Z., Václavík, M.: Cam Mechanisms, vol. 14. Elsevier Science Limited (1993)
12. Kuzenkov, V.V., Samoilova, M.V., Tarabarin, V.B., Timofeev, G.A., Umnov, N.V.: Theory of Mechanisms and Machines. Handbook on the Course Project, 2, Remastered edn. Bauman Moscow State Technical University. Publishing House, Moscow (2012)
13. Larson, J., Cheng, H.H.: Object-oriented cam design through the internet. J. Intell. Manuf. **11**(6), 515–534 (2000). https://doi.org/10.1023/A:1026548305291
14. Leonov, I.V., Baryshnikova, O.O., Kuzenkov, V.V., Sinitsin, V.V., Tarabarin, V.B.: Usage of Mathcad system while developing of course projects and hometasks of theory of mechanisms and machines. Bauman Moscow State Technical University, Moscow (2004). (in Russian)
15. Makaroff, E.G.: Engineering Calculations in Mathcad 15. Piter, St. Petersburg (2011). (in Russian)
16. Petropoulou, A., Dimopoulos, S., Mourtzis, D., Chondros, T.G.: A computer aided method for cam profile design, pp. 369–376. Springer, Netherlands (2009). https://doi.org/10.1007/978-1-4020-8915-2_45
17. Pylaev, B.V.: Method of flat cams profiling. Herald of the Voronezh State Agrarian University (1), 78–81 (2010). (in Russian)

18. Satyanarayana, B., Rao, P.N., Tewari, N.K.: Machining of plate cam profiles on CNC machine tools using a highly integrated part programming system. Int. J. Adv. Manuf. Technol. **3**(4), 105–125 (1988). https://doi.org/10.1007/BF02601837
19. Vukolov, A.: F. Reuleaux, F. Wittenbauer: their influence on evolution of applied mechanics in Russia at the beginnings of XXth century, pp. 315–322. Springer, Cham (2017). https://doi.org/10.1007/978-3-319-44156-6_32
20. Vukolov, A., Golovin, A.: A.N. Krylov: the pioneer of photographic non-invasive measurement methods in Russian science, pp. 903–911. Springer, Cham (2015). https://doi.org/10.1007/978-3-319-09411-3_94

On the Use of Gröbner Bases in a Robotics Course

Manfred Husty[1(✉)] and Paul Zsombor-Murray[2]

[1] Unit Geometry and CAD, University of Innsbruck, Innsbruck, Austria
Manfred.Husty@uibk.ac.at
[2] CIM, McGill University, Montreal, Canada
paul@cim.mcgill.ca

Abstract. Most robotic mechanical systems consist of rigid links connected via revolute or prismatic joints. Therefore their kinematic behaviour can be described by nonlinear algebraic (polynomial) equations. These equations are called constraint equations and can be used to solve the most important kinematic problems like forward and inverse kinematics or singularities. Whereas methods from algebraic geometry to treat these equations are used in scientific papers, these often very elegant methods are rarely used in engineering courses. This paper reports on some of these methods and how they were used in a robotics course.

1 Introduction

The use of Gröbner bases in an engineering course needs some prerequisites and a certain sequence of contents. the first step is to introduce the rational representation of the Euclidean displacement, which is done using Study's kinematic mapping. This generally need no more than one lecture, because the students know already other representations like Euler angles, Rodrigues parameters or Euler parameters. The second step is a brief introduction into Gröbner bases and the notions of algebraic geometry which are linked to this tool. A very rough introduction to this mathematical concept done without any proof, needs no more than one lecture. Then the application of this tool can start. Within this paper due to the limitation of space only the basic ideas can be described and two out of many application will be shown. The paper is organized as follows: In Sect. 2 a brief introduction to the kinematic image space is given. Section 3 contains the basic definitions from algebraic geometry and Sect. 4 shows two examples where the previously developed concepts have been applied. These are: the general equation of the coupler curve of a four-bar mechanism and the direct kinematic of the Stewart platform. The paper finishes with a conclusion.

2 Rational Representation of Euclidean Displacements, Kinematic Mapping

This brief introduction to the concept of rational representation of Euclidean displacements is taken from our paper [4]. Euclidean rigid body displacements

J. C. García-Prada and C. Castejón (Eds.): *New Trends in Educational Activity in the Field of Mechanism and Machine Theory*, MMS 64, pp. 20–28, 2019.
https://doi.org/10.1007/978-3-030-00108-7_3

are often described by homogeneous 4×4 matrices $\mathbf{M}$, that act on a point $\mathbf{x}$ located in a moving frame according to $\mathbf{x}' = \mathbf{M}\mathbf{x}$. $\mathbf{x}'$ is the image point in a base frame, the lower right 3×3 sub matrix of $\mathbf{M}$ is a proper orthogonal matrix encoding the orientation of the moving fame with respect to the base frame and the first column of $\mathbf{M}$ contains the vector connecting the origins of both frames therefore representing the translational part of the transformation. Using Study's kinematic mapping κ (see [3,7]), the displacement given by $\mathbf{M}$ is mapped to a point $\mathbf{d} = [x_0, x_1, x_2, x_3, y_0, y_1, y_2, y_3]^T$ in a seven dimensional projective space P^7. These point coordinates are called the *Study parameters* of the displacement. They fulfill the quadratic condition

$$x_0y_0 + x_1y_1 + x_2y_2 + x_3y_3 = 0, \tag{1.1}$$

which is called *Study condition*. Its zero set is the *Study quadric* $S_6^2 \subset P^7$.

In the inverse kinematic mapping a point on S_6^2 minus the *exceptional three space* $\mathcal{E}\colon x_0 = x_1 = x_2 = x_3 = 0$ yields the matrix

$$\mathbf{M} := \kappa^{-1}(\mathbf{d}) = \frac{1}{\Delta}\begin{bmatrix} 1 & 0 & 0 & 0 \\ t_1 & x_0^2 + x_1^2 - x_3^2 - x_2^2 & -2x_0x_3 + 2x_2x_1 & 2x_3x_1 + 2x_0x_2 \\ t_2 & 2x_2x_1 + 2x_0x_3 & x_0^2 + x_2^2 - x_1^2 - x_3^2 & -2x_0x_1 + 2x_3x_2 \\ t_3 & -2x_0x_2 + 2x_3x_1 & 2x_3x_2 + 2x_0x_1 & x_0^2 + x_3^2 - x_2^2 - x_1^2 \end{bmatrix} \tag{1.2}$$

where $\Delta = x_0^2 + x_1^2 + x_2^2 + x_3^2$ and

$$\begin{aligned} t_1 &= 2x_0y_1 - 2y_0x_1 - 2y_2x_3 + 2y_3x_2, \\ t_2 &= 2x_0y_2 - 2y_0x_2 - 2y_3x_1 + 2y_1x_3, \\ t_3 &= 2x_0y_3 - 2y_0x_3 - 2y_1x_2 + 2y_2x_1. \end{aligned} \tag{1.3}$$

Properties of kinematic chains are described with respect to (arbitrarily chosen) base and end effector coordinate frames. Possible locations of the end-effector with respect to the base correspond to algebraic varieties described by sets of polynomial equations in $\mathbb{P}^7$. The representation of planar displacements can be obtained by setting $x_1 = x_2 = y_0 = y_3 = 0$ and taking the upper left 3×3 sub-matrix of Eq. 1.2.

3 Algebraic Geometry Basics

In this introduction only the most important definitions and theorems are given. It is based on the book *"Ideals, Varieties, and Algorithms"* by *David Cox, John Little* and *Donal O'Shea* [2], which is an excellent introduction to algebraic geometry. More detailed descriptions and also the proofs for the theorems can be found there.

Examples were computed using *Maple 18* with its package `Groebner`. This package contains the low-level commands which were enough to compute all examples of the course, the `PolynomialIdeals` package is newer and contains the more sophisticated commands. There are also other software packages like e.g. *Singular*, *Macaulay 2* or *Mathematica* for such computations.

In the following all algebraic equations are polynomials in the ring $\mathbf{K}[\underline{x}] = \mathbf{K}[x_1, \ldots, x_n]$ where $\mathbf{K}$ is a field like $\mathbb{Q}$ or $\mathbb{C}$.

Ideals and Affine Varieties

At first polynomial ideals are defined which are the basic objects for everything else.

Definition 1. A set $I \subseteq \mathbf{K}[\underline{x}]$ is called an **ideal** if the following conditions are fulfilled:

- $\forall f, g \in I : f + g \in I$
- $\forall f \in I$ and $\forall h \in \mathbf{K}[\underline{x}] : hf \in I$

It follows that almost all ideals are infinite sets of polynomials and cannot be written down as a whole. The sole exception is the ideal $\{0\}$ which is also a proper subset resp. subideal of all other ideals because 0 is contained in every ideal. There is also an ideal which is a proper superset resp. superideal of them, namely the ideal which contains the constant polynomial 1 and with it all polynomials of $\mathbf{K}[\underline{x}]$.

Using Definition 1 it is possible now to define the ideal generated by a set of given polynomials $f_1, \ldots, f_s$.

Definition 2. Let $f_1, \ldots, f_s$ be polynomials in $\mathbf{K}[\underline{x}]$. Then the set

$$\langle f_1, \ldots, f_s \rangle = \{g \in \mathbf{K}[\underline{x}] : g = \sum_{i=1}^{s} h_i f_i \ \textit{and} \ h_1, \ldots, h_s \in \mathbf{K}[\underline{x}]\}$$

is the **ideal generated by** $f_1, \ldots, f_s$.

In kinematics of mechanical systems one is much more interested in the affine variety that is associated with an ideal. Its definition is given below

Definition 3. For a given ideal $I = \langle f_1, \ldots, f_s \rangle \subseteq \mathbf{K}[\underline{x}]$ the set

$$\mathbf{V}(I) = \{(a_1, \ldots, a_n) \in \mathbf{K}^n : f_i(a_1, \ldots, a_n) = 0 \ \textit{for all} \ 1 \leq i \leq s\} \subseteq \mathbf{K}^n$$

is called the **affine variety** of the ideal I.

The affine varieties are therefore the zero sets of polynomial equations. They describe the points which fulfill the kinematic equations. The objects of interest have been defined in a quite abstract way and now one needs methods of find their solution sets. The general situation in kinematics is as follows: the kinematic links together with the joints connecting them define a kinematic chain that imposes constraints to the end effector of the chain. The constraints can be expressed in polynomial form. A set of constraints forms a set of polynomials and generates, according to the definition above, an ideal. The solution set corresponding to the set of constraints is the affine variety belonging to this ideal.

As one has to be aware that the polynomials in the constraint equations are generally multivariate and solution algorithms depend heavily on multivariate divisions one needs at first some definitions on the term ordering in multivariate polynomials. The most important term ordering is given as an example.

Definition 4 (Lexicographic Order). Let $\alpha = (\alpha_1, \ldots, \alpha_n)$ and $\beta = (\beta_1, \ldots, \beta_n)$ be elements of $\mathbb{Z}^n_{\geq 0}$. We define $\alpha >_{lex} \beta$ if the leftmost nonzero entry of the vector-difference $\alpha - \beta \in \mathbb{Z}^n$ is positive.

In Maple the keyword for this ordering is `plex`, e.g. a possible lexicographic termorder for polynomials containing the unknowns $\{x_1, x_2, x_3\}$ could be `plex(x[3], x[1], x[2])`. There exist other termorders like Graded Lex Order (Maple: `tdeg`) or Graded Reverse Lex Order. There is one more important termorder which is used in elimination processes:

Definition 5. lexdeg($L[1], ..., L[k]$) is an elimination order. The L[i] must be disjoint lists of variables. The variables in $L[i]$ are eliminated from the subsequent variables for all $i = 1..k - 1$.

This order is often used to eliminate variables when one does not want to compute an entire "plex" basis. It is equivalent to a product order that uses"tdeg" on each L[i]. The last ingredients necessary to define Gröbner bases is the notion of leading monomial, leading coefficient and leading term: A Gröbner basis is a very special generating set of the ideal I with some nice properties. Such a basis is not unique due to the fact that adding another polynomial to the generators does not change this property. Another reason for non-uniqueness is the fact that a monomial order has to be chosen first. One of the nicest properties of a Gröbner basis computed with a lexicographic term order ("plex" in Maple) is that it has the so called elimination property. To explain this property consider the following simple example:

Example 1. *The ideal* $I = \langle x_1^2 x_2 + x_1 x_2^2 + x_2^2 x_3, x_1 x_2 - x_3, x_2 x_3^2 - 1 \rangle$ *is given with the Gröbner basis wrt.* `plex(x[1],x[2],x[3])`

$$G = \langle x_3^7 + x_3^2 + 1, x_3^5 + x_2 + 1, -x_3^3 + x_1 \rangle$$

The first entry of G is a univariate polynomial in x_3 and the remaining two polynomials contain the other two unknowns linearly, which means after solving the first polynomial for X_3 the back substitution becomes linear in the other variables. The univariate will always be in the unknown which is last in the sequence plex$(x_i, x_j, x_k), i, j, k = 1, 2, 3$. Note that the back substitution is not necessarily linear, depending on the geometric structure of the set of starting equations.

4 Examples

4.1 Four-Bar Coupler Curve

In this example a planar four-bar is given with base points A_1, A_2, coupler points a_1, a_2 and the two arm lengths r_1, r_2 (Fig. 1). The question is to compute the coupler curve of a point $E(x, y)$ of the coupler system $(a_1; x, y)$. There are classical methods to do this, e.g. using complex numbers [9, pp. 66–69] or

Cartesian coordinates [8, pp. 178–180]. Here we use the kinematic mapping and the Gröbner basis to obtain the equation of the coupler curve. In the Maple worksheet below the coupler curve for the design values of Fig. 1 is computed ($A_1(0,0), a_1(0,0), A_2(5,0), a_2(3,0), r_1 = 3, r_2 = 4, E(2,1)$). T defines the general transformation matrix between the coupler system and the base. The two equations $gl1$ and $gl2$ say that the distances between points A_i and a_i are constant. $gl3$ and $gl4$ are equations coming from the transformation of point E to the base. Therefore one has four equations $gli, i = 1,..4$. As the motion coordinates are homogeneous a homogenizing equation is added to a system of five polynomial equations. The Maple command `lexdeg([x0, y1, y2, x3],[x, y])` computes a Gröbner basis in which the first equation contains only the variables x, y and yields therefore the sought equation of the coupler curve. Note that the same procedure can be run without assigning values to the design variables, leading to the general equation of a coupler curve.

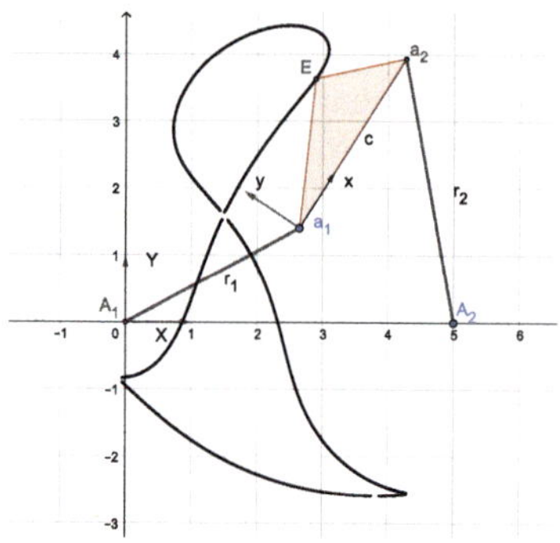

Fig. 1. Four-bar with coupler curve

Maple worksheet for the computation of the coupler curve

```
> T:= Matrix(3,3,{(1,1)=1,(1,2)=0,(1,3)=0,
> (2,1)=(-2*x0*y1+2*x3*y2)/(x0^2+x3^2),(2,2)=(x0^2-x3^2)/(x0^2+x3^2),
> (2,3)=-2*x0*x3/(x0^2+x3^2),(3,1)=(-2*x0*y2-2*x3*y1)/(x0^2+x3^2),
> (3,2)=2*x0*x3/(x0^2+x3^2),(3,3)=(x0^2-x3^2)/(x0^2+x3^2)}):
> A1:=Vector([1,0,0]):
> A2:=Vector([1,5,0]):
> a1:=Vector([1,0,0]):
> a2:=Vector([1,3,0]):
> gl1:=op(3,factor(numer(Transpose(A1-T.a1).(A1-T.a1)-3^2))):
> gl2:=op(3,factor(numer(Transpose(A2-T.a2).(A2-T.a2)-4^2))):
> b1:=Vector([1,1,2]):
> B:=Trans.b1:
> gl3:=numer(normal(B[2]-x)):
> gl4:=numer(normal(B[3]-y)):
> F:=[gl1,gl2,gl3,gl4,x0^2+x3^2-1]:
> bas:=Groebner[Basis](FF,lexdeg([x0,y1,y2,x3],[x,y])):
```

The first entry in the Gröbner basis *bas* is the implicit equation of the coupler curve:

$$\begin{aligned} &9\,x^6+27\,x^4y^2+27\,x^2y^4+9\,y^6-150\,x^5-30\,x^4y-300\,x^3y^2-60\,x^2y^3\\ &-150\,xy^4-30\,y^5+722\,x^4+300\,x^3y+844\,x^2y^2+300\,xy^3+122\,y^4-570\,x^3\\ &-330\,yx^2-570\,xy^2-330\,y^3-1151\,x^2-600\,yx+49\,y^2+120\,x+1320\,y+800=0 \end{aligned} \tag{1.4}$$

Remark 1. *Fig. 1 was generated with Geogebra [1], a free dynamical 2d and 3d software package. The students can draw the four-bar linkage and animate it. The coupler curve equation can be generated in Geogebra using either the animation and the* `Locus` *command which might give an incomplete curve or the* `ImplicitCurve(f<xy>)` *command using the Eq. 1.2. This will produce the complete coupler curve. It is also possible to transfer the general coupler curve equation into Geogebra and play dynamically with the six design parameters* $(A_1, a_1, r_1, r_2, x, y)$ *of the four-bar linkage and the coupler point. With this the students can study the variety of different coupler curves. With the coupler curve equation also the double points and the focal circle can be found easily.*

4.2 Direct Kinematics of the Stewart Platform

In [5] the first algorithm to solve the direct kinematics (DK) of general Stewart-Gough platform (SGP) manipulators was presented. Using the kinematic constraint equations that comprise the given distances between the base and platform anchor points it was shown that after a sophisticated elimination process the final univariate polynomial in one of the unknowns could be computed. After numerical solving of this polynomial in a relatively complicated back substitution the remaining pos variables were found. This process was to complicated to be taught in a basic robotics course. The Gröbner basis algorithm which is now implemented in most of the symbolc manipulation software systems simplifies this task such that the DK of SGP can be taught even in a first robotics course. Due to lack of space a commented Maple worksheet is presented to show how the algorithm can be performed. Some of the display is truncated to fit in the paper. The design parameters used in this example are taken from the lower SGP of the hexapod telescope [6].

Maple worksheet: DK of the hexapod telescope manipulator.

T is the algebraic Transformation matrix between the platform and the base of the manipulator (see Fig. 2)

```
> T:=Matrix(4,4,{(1,1)=1,(1,2)=0,(1,3)=0,(1,4)= 0,
> (2,1) = (-2*x0*y1+2*x1*y0-2*x2*y3+2*x3*y2)/(x0^2+x1^2+x2^2+x3^2),
> (2,2) = (x0^2+x1^2-x2^2-x3^2)/(x0^2+x1^2+x2^2+x3^2),
> (2,3) = (-2*x0*x3+2*x1*x2)/(x0^2+x1^2+x2^2+x3^2), (2, 4) = (2*x0*x2+2*x1*x3)/(x0^2+x1^2+x2^2+x3^2),
> (3,1) = (-2*x0*y2+2*x1*y3+2*x2*y0-2*x3*y1)/(x0^2+x1^2+x2^2+x3^2),
> (3,2) = (2*x0*x3+2*x1*x2)/(x0^2+x1^2+x2^2+x3^2),
> (3,3) = (x0^2-x1^2+x2^2-x3^2)/(x0^2+x1^2+x2^2+x3^2), (3,4)= (-2*x0*x1+2*x2*x3)/(x0^2+x1^2+x2^2+x3^2),
> (4,1) = (-2*x0*y3-2*x1*y2+2*x2*y1+2*x3*y0)/(x0^2+x1^2+x2^2+x3^2),
> (4,2) = (-2*x0*x2+2*x1*x3)/(x0^2+x1^2+x2^2+x3^2), (4,3) = (2*x0*x1+2*x2*x3)/(x0^2+x1^2+x2^2+x3^2),
> (4,4) = (x0^2-x1^2-x2^2+x3^2)/(x0^2+x1^2+x2^2+x3^2)}):
```

A_1 are the coordinates of the base anchor points (centers of the spherical joints), a_1 are those of the platform anchor points in platform coordinate system and B_1 are the coordinates of b_i measured in the base system.

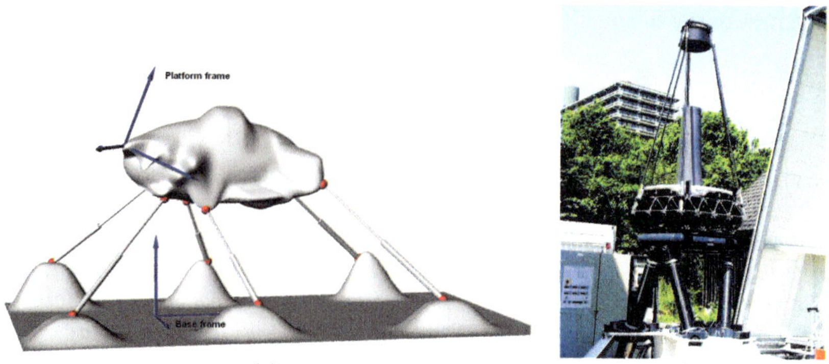

(a) (b)

Fig. 2. (a) Stewart-Gough Platform Manipulator (b) Hexapod telescope

```
> for i from 1 to 6 do
> A||i:=Vector([1,a1||i,a2||i,a3||i]);
> b||i:=Vector([1,b1||i,b2||i,b3||i]);
> B||i:=T.b||i; od:
```

h_i are the distance constraint polynomials, r_i the given leg lengths and by adding 4 time the square of the Study quadric (Eq. 1.1) h_i become quadratic. This is allowed because of Definition 1.

```
> for i from 1 to 6 do
> h||i:=op(2,factor(numer(Transpose(A||i-B||i).
> (A||i-B||i)-r||i^2)+4*StudyQuad(stud)^2));
> od:
> a16:=a11:a15:=a12:a14:=a13:a26:=-a21:a25:=-a22:a24:=-a23:
> b16:=b11:b15:=b12:b14:=b13:b26:=-b21:b25:=-b22:b24:=-b23:
> a31:=0:a32:=0:a33:=0:a34:=0:a35:=0:a36:=0:b31:=0:b32:=0:
> b33:=0:b34:=0:b35:=0:b36:=0:
```

here the design parameters of the manipulator are set

```
> a11:=752:a12:=-139:a13:=-613:a21:=274:a22:=788:a23:=514:
> b11:=426:b12:=235:b13:=-661:b21:=517:b22:=627:b23:=110:
> r1:=730:r2:=727:r3:=719:r4:=721:r5:=737:r6:=741:
```

F is the definition of the ideal consisting of 8 polynomials in 8 unknowns, $x0 = 1$ is used as normalization of the homogeneous variable and again Definition 1 is applied, because differences of constraint equations are linear in the unknowns y_i (see [5])

```
> F:=[x0-1,h1,h2-h1,h3-h1,h4-h1,h5-h1,h6-h1,StudyQuad(stud)];
```

$$F := [x0 - 1, -367575\, x0^2 + 1088240\, x0\, x3 + 1304\, x0\, y1 - 972\, x0\, y2 + \ldots]$$

It makes sense to compute at first a total degree (tdeg) basis, because this usually is faster than a plex basis

```
> tdegBas:=Groebner[Basis](F,tdeg(y0,y1,y2,y3,x0,x1,x2,x3)):
```

From the tdeg basis a plex basis is computed in less than 1s on a notebook computer. The sequence of the unknown is important! The basis consists of eight polynomials.

```
> plexBas:=Groebner[Basis](tdegBas,plex(y0,y1,y2,y3,
x0,x3,x2,x1)):
```

Now the numerical part starts, one can set the accuracy to any digit wanted

```
> Digits:=40:
```

the first entry of the basis is the univariate which is solved for its unknown

```
> Lx1:=fsolve(op(1,plexBas),x1);
```

$$Lx1 := -0.49289752572694..., -.492573813561538..., \ldots$$

the first solution is inserted in the remaining 7 polynomials of the plex basis, which are then solve linearly! for the remaining variables

```
> LX1:=eval(plexBas,x1=Lx1[1]):
> fsolve({LX1[2],LX1[3],LX1[4],LX1[5],LX1[6],LX1[7],LX1[8]},
{x0,x2,x3,y0,y1,y2,y3});
```

$$\{x0 = 1.0, x2 = -0.271.., x3 = -0.00084.., y0 = 5.0159.., y1 = 109.1718.., y2 = -179.753..\}$$

5 Conclusions

The most universally useful aspect of applying Gröbner bases as a pedagogical tool is that with it the student has freedom to explore all problems that can be formulated as simultaneous polynomial equations. Thus one obtains clear insight into underlying structure and how a particular apparently novel unrelated problem is intimately but unexpectedly connected to a host of situations that have become familiar through previous study. This freedom is provided because the tool is so convenient and intuitively easy to use.

References

1. Geogebra. www-geogebra.org
2. Cox, D.A., Little, J., O'Shea, D.: Ideals, Varieties, and Algorithms: An Introduction to Computational Algebraic Geometry and Commutative Algebra, 3rd edn. Springer, New York (2007)
3. Husty, M., Pfurner, M., Schröcker, H.P.: Algebraic methods in mechanism analysis and synthesis. Robotica **25**(6), 661–675 (2007)
4. Husty, M., Schadlbauer, J., Zsombor-Murray, P.: Cable-driven parallel robots. In: A New Approach to the Direct Geometrico-Static Problem of Cable Suspended Robots Using Kinematic Mapping, Mechanism and Machine Science, no. 53, pp. 97–105. Springer International Publisher (2018)
5. Husty, M.L.: An algorithm for solving the direct kinematics of general Stewart-Gough platforms. Mech. Mach. Theor. **31**(4), 365–379 (1996)
6. Husty, M.L., Eberharter, H.: Kinematic analysis of the hexapod telescope. In: Park F., Iurascu, C. (eds.) Proceedings of the 2nd Workshop on Computing Kinematics, pp. 269–278. Seoul, Korea (2001)

7. Study, E.: Geometrie der Dynamen. B. G. Teubner, Leipzig (1903)
8. Wieleitner, H.: Spezielle ebene Kurven. Sammlung Schubert LVI. G.J. Göschnesche Verlagshandlung, Leipzig (1908)
9. Wunderlich, W.: Ebene Kinematik. No. 447/447a in B.I. Hochschultaschenbücher. Bibiliographisches Institut, Mannheim (1970)

New Educational Methodology to Study the Involute Tooth Profile

Olga Egorova(✉), Gennady Timofeev, and Georgy Shashurin

Bauman Moscow State Technical University,
107005 2-ya Baumanskaya, 5, Moscow, Russia
tmm-olgaegorova@yandex.ru

Abstract. The paper presents a new educational methodology to study of spur (cylindrical) involute gears with the help of laboratory equipment and its three dimensional (3D) computer models. During the laboratory practice at the «Theory of Mechanisms and Machines» (TMM) Department at Bauman Moscow State Technical University (Russia) the students learn the basic theory and sketch (by tracing) the involute tooth profile of a spur gear that is formed by using a generating method and a shaper cutter as a cutting tool. Sketching is carried out with use of standard laboratory equipment, including TMM-47 device and its 3D-model, as well as others. Devices and models imitate an involute gearing and give an evident idea of the metal-removal process and formation of involute tooth profile. The proposed methodology is related to the information-and-developing technologies and aimed at the best comprehensibility of discipline, popularization and dissemination of scientific knowledge with an increasing effect in case of open access to 3D-documents hosted on Internet network.

Keywords: Involute spur gears · Shaper-cutter · 3D modeling
Theory of Mechanisms and Machines (TMM) · Generating method

1 Introduction

Modern engineer of any specialty needs knowledge connected with the gear design as they have the largest area of application among a big variety of the existing mechanisms and machines. Gear design is considered to be one of the most important and complicated fields of Mechanical Engineering. The most important characteristic of gearing is the gear ratio. During the meshing of one pair of teeth the value of gear ratio depends mostly on a profile of a tooth. This fact follows from the basic theorem of correct mesh (Willis's theorem).

Robert Willis (1800–1875), an English academic (Fig. 1a), in 1841 published his "Principles of Mechanism" [1] (Fig. 1b) - one of the first classic and seminal works devoted to Mechanism and Machine Science (MMS). Besides, in 1837, Professor Willis read to the British Association a valuable paper «Teeth of Wheels» as well as many works on medieval architecture and the mechanical construction of English cathedrals, notable for his incisive decompositions of these structures' functional and decorative aspects [2]. He was the first Cambridge professor to win widespread

J. C. García-Prada and C. Castejón (Eds.): *New Trends in Educational Activity in the Field of Mechanism and Machine Theory*, MMS 64, pp. 29–37, 2019.
https://doi.org/10.1007/978-3-030-00108-7_4

recognition as a Mechanical Engineer. In his scientific works Willis defined the concept of mechanism; updated the theory of mechanisms (TMM); discovered and developed many new theories, including the basic theorem for correct mesh, later named after him (Willis's theorem), which in modern interpretation reads as follows: «Common normal to the contacting tooth profiles at the moment of engagement divides the center line of the cogwheels into the parts inversely proportional to the angular velocities».

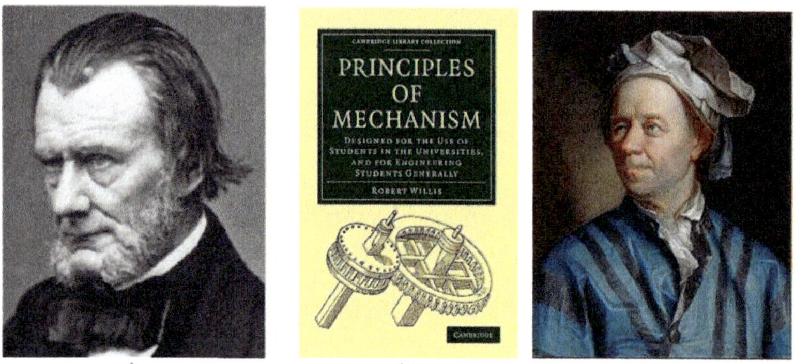

a) b) c)

Fig. 1. (a) Robert Willis [2]; (b) cover of "Principles of Mechanism" [1], (c) Leonhard Euler [4]

However, almost a century before Robert Willis had published his "Principles of Mechanism", in the middle of the 18th century the involute tooth profile was originally designed by Leonhard Euler (1707–1783), the great mathematician and scientist of Swiss origin (Fig. 1c) [3–9]. They say, in connection with the design of water turbines Euler developed optimal profiles for teeth in cogwheels that transmit motion with a minimum of resistance and noise [10]. It is definitely known that in 1754 Euler published several papers on cogwheels at the Berlin Academy. Later, in 1765, he returned to this topic in his work «Supplementum de figura dentium rotarum» [20] and proved the ideal teeth should be formed by the involute of a circle. In this, the teeth have a common tangent, and the cogwheel does not vibrate, the energy is not wasted on noise, and the costs become minimal. The technical realization of this design took shape only later in what is called now the involute gearing that became the most commonly used [4, 9, 10]. Euler was the first scientist to study the involute gearing and his ideas led to the geometric equations, now usually called after him as the Euler–Savary equations.

Currently more than 90% of all gears in the world are involute. Why this happened? The main reasons are: the involute tooth profile provides a constant gear ratio and allows small Center Distance changes.

It is obvious, all studies of the involute tooth profile are very important. However, along with the development of "TMM" discipline, professors and lecturers always implement new textbooks, new manuals, new equipment and devices, new educational methodologies [15]. Today, in the 21st century, modern computers and automated programs help in solution of many TMM tasks, including both synthesis and analysis. 3D modeling gives us new opportunities, involving the students into the process of studying by modeling laboratory equipment, and scaled models of mechanisms in particular [15–17]. New educational methodologies used at Bauman Moscow State Technical University (www.bmstu.ru) force students to study Theory of Mechanisms and Machines through 3D modeling. As a result, they learn a lot in the field of TMM along with implementation of their own experience and knowledge of previous courses and disciplines.

2 TMM-47 Device: Tracing Involute Tooth Profile

2.1 Involute Tooth Profile

According to the Gear Nomenclature [11] «The involute tooth profile is the most general gear tooth form. The involute curve is the trace that the end of a taut string produces as it is unwound from a cylinder, and the gear tooth whose cross section is the involute curve is called the involute tooth profile. When two gears with involute profiles mesh together the contact point moves along the common tangent of the two base circles. Also, because the contact surface is always perpendicular to the common tangent, it has the feature of having the direction of the acting force along the same common tangent. That is to say, with involute shaped gears, the force is transmitted in one direction securely from the beginning to the end of the mesh. The characteristics of the involute gears include that even with some errors in the Center Distance, they can still mesh and they are easy to generate». It is also mentioned in the «Elements of metric gear technology», [12] that involute gears have three major advantages:

1. Conjugate action is independent of changes in Center Distance;
2. The form of the basic rack tooth is straight-sided, and therefore is relatively simple and can be accurately made;
3. One cutter can generate all gear tooth numbers of the same pitch.

High speed and/or high power gear trains require transmission at constant angular velocities in order to avoid severe dynamic problems. Thus, a primary requirement of gears is the constancy of angular velocities or proportionality of position transmission. Involute gears are perfect from this point of view.

2.2 Undercut: Pointed Tip of Tooth

The gears with the correct involute tooth profiles transmit power efficiently and correctly. When a pair of gears meshes, the distance between the centers of the meshing gears' shafts is called its Center Distance [11, 12]. It is calculated as the half of the sum of two gears' pitch diameters. Formula for Center Distance looks like:

$$a_w = (d1 + d2) / 2.$$

However, the smaller number of teeth can achieve a more reliable transmission when the Center Distance is fixed. But small number of teeth could produce the problem of undercut that is associated with the phenomenon of cutting the root of a gear tooth deeper than the involute tooth curve. This can happen when there is interference between the cutting tool and the gear. The undercut of pinion teeth is undesirable because of losses of strength, contact ratio and smoothness of action, but when it is small then in many applications its adverse effects can be neglected. For the very small numbers of teeth, such as ten and smaller, and for high-precision applications, the undercut is an important problem and should be avoided [11, 12].

When the undercut is large, the root of the gear becomes narrower and the tooth becomes weak in its bending strength, and the load capacity of the tooth is reduced and the length of the line of action may reduce. The undercut also removes some of the useful involute adjacent to the base circle. Thus, the transmission capacity of a gear-set is substantially reduced. Beside, the undercut may cause dulling wear of the shaper cutter on its corners. For all these reasons the undercut is undesirable. The general conditions to avoid the undercut have been proposed by Russian (Soviet) scientists Professors Vladimir A. Gavrilenko and Faydor L. Litvin.

The undercut occurs when it is needed to cut a gear with a small number of teeth (for a standard gear with a pressure angle α equal to 20°, number of teeth z equal to 17 or less). In this, the tip of the cutting tool extends beyond the interference point [11].

To prevent the undercut and to provide a strength balance between a pinion and a gear without changing the number of teeth, a method called profile shifting (or profile modification) is used. Profile shifting method is also used to accommodate certain Center Distances that are considerably afar from the theoretical Center Distance for a standard gear-set.

Profile shifting means the depth of the cutting tool is changed. To prevent the undercut a positive profile shifting is used. In this, the root of a gear tooth becomes thicker that is good. But there must be a limit on the magnitude of positive profile shifting, because when it is excessive, the tip of a gear tooth becomes pointed [11].

2.3 Gears Shaping: Gear Shaper Cutter

Gear shaping is one of the gear generating methods. In this process gear teeth are accurately sized and shaped by cutting them by a multipoint cutting tool: gear shaper cutter, rack type cutter, gear hob, etc. The name shaper relates to the fact that the cutter engages the part on the forward stroke and pulls away from the part on the return stroke, just like the clapper box on a planer shaper. Gear shaper cutter (Fig. 2) is the gear cutting tool with cutting edges on its outer surface. The both – the gear blank and the gear shaper cutter are mounted on the gear cutting machine. Then a symmetrical motion of rotation and reciprocating generates the gear teeth.

Gear shaper cutter is used for roughing and finishing of gears with internal and external teeth, if a small gap for the exit of the tool does not allow the use of any other cutter. However, in most cases the gear shaping by a gear shaper cutter is inferior to gear milling or worm milling cutters in terms of productivity and accuracy. However, the main advantage of the first is the possibility to cut the involute profile small-toothed wheels.

Producing the contour of the shaper cutter [13] with sufficient for practical calculations, the accuracy is considered to be involute.

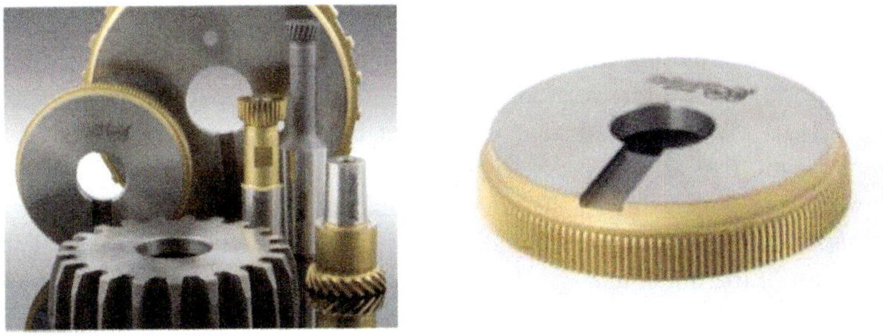

Fig. 2. Gear shaper cutters produced by Star-SU. (http://www.star-su.com/cutting-tools/gear-cutting-tools/gear-shapers)

2.4 TMM-47 Device

At the «Theory of Mechanisms and Machines» Department at Bauman Moscow State Technical University the main task for students at laboratory practice connected with gear design is to study the above mentioned advantages of involute tooth profile, the undercut and the pointed tip of a gear tooth when a gear has very small number of teeth are to be produced [14]. Study of spur involute gears goes on with the help of the laboratory equipment. Tracing the involute tooth profile personally during laboratory practice with the help of TMM-47 device gives opportunity to each student getting his own experience in studying TMM, and in gear design in particular.

The TMM-47 device (Fig. 3) imitates a gearing and is intended for receiving an evident idea of the metal-removal process (cutting) and forming of involute tooth profile of a spur gear that is formed by generating method using a gear shaper cutter as a cutting tool.

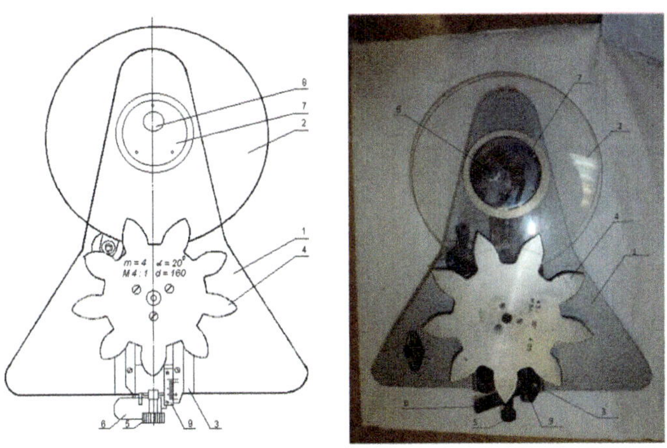

Fig. 3. Scheme and a photo of the TMM-47 device

The TMM-47 device consists of a transparent plastic disk 2, fixed on a cast basis 1, and a metal gear 4 imitating a shaper cutter as a cutting tool. The sizes of the shaper cutter correspond to his initial section. The disk 2 and the shaper cutter 4 are produced in a scale M 4:1. A paper disk that imitates a blank is fixed on 4 needles on the transparent plastic disk 2 and pressed by a washer 7 and a screw 8. Flexible transmission provides a possibility of the simultaneous rotation of the blank (paper disk 2) and the shaper cutter 3 with the angular velocities associated to gear ratio. The interrelated rotation of the blank and the shaper cutter for a limited angle could be made by a special mechanism by pressing the key 6. If consistently after each keystroke to trace by a pencil on the paper disk 2 all the teeth of the shaper cutter, then we could see a number of the contours that are shifted among them in relative motion. The envelope to these contours gives the involute profile of the gear tooth.

This method is called the generating method, at which the cutting tool and the blank due to kinematical chain of the TMM-47 device perform two movements – «cutting» and generating (under generating is defined as the relative movement of the blank and the tool which meets machining engagement). The tooth profile is obtained as the envelope of the successive provisions of the tool cutting edge.

2.5 Tracing Involute Tooth Profile

A gear shaper cutter can be installed at a different distance from the center of the blank (a_w – Center Distance), making it possible to cut gears with or without a profile modification. The shaper cutter installation in the radial direction towards the center of the blank is carried out with the help of screw mechanism 7 (Fig. 3).

The value of the profile shifting is determined by the mutual position of the two scratches on the housing and a movable scale attached to the sled 9. If the extreme risks of the movable scale coincides with the two marks on the housing means the profile shifting is zero. Positive or negative profile shifting could be controlled easily according to the scratches with the «+» or «−» sign.

Before tracing the profile of the cylindrical involute wheels with and without offset shaping cutter (on the device TMM-47), a student calculates the parameters of these wheels. The results of tracing the teeth of a shaper cutter on the device TMM-47 are presented on Fig. 4.

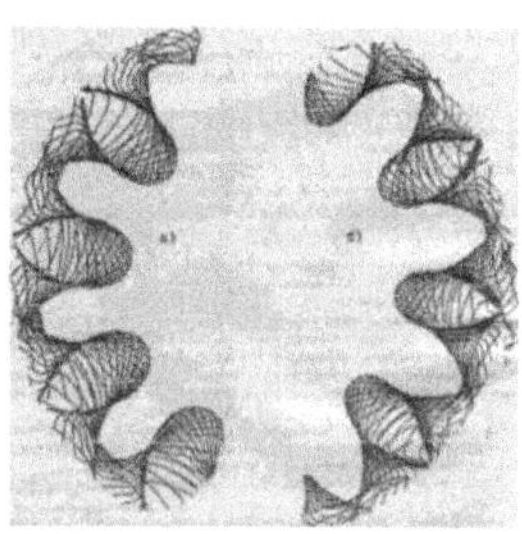

Fig. 4. Paper disk (blank) with involute profile teeth with (b) or without any modification (a)

3 3D Modeling of Laboratory Equipment

Today, the widespread use of IT technologies supports a lot of attempts to move a complex educational process in virtual working environment. Besides, modern computers and automated programs help in solution of many TMM tasks, including both synthesis and analysis [19].

New educational methodology based on a wide use of multimedia possibilities became in a great demand. Lectures and workshops are going on accompanied by a computer presentation along with demonstration of mechanisms digital photos, virtual copies of scaled mechanism's model and laboratory equipment as well as video movies, 3D animation, etc.

Moreover, attracting to 3D modeling and searching related webpage to expand needed information through Internet gives students more profound knowledge. Before a student starts any creation of a 3D document he needs to identify and describe the object. This work helps student to get practical skills as well as to learn a lot on Mechanical Engineering, History of TMM, Machine and Mechanism Science and other related fields [9, 15–19].

Modern education prefers teaching based on creating 3D models (Fig. 5), rather than the more traditional methods based on use of card, balsa, paint and glue models or even on manufacturing of wooden or metal models [16]. Thus, modern methodology focuses on the process of developing digital and computer models [9] that could be also used to explore and illustrate in future the possibilities of real mechanism or machine.

a) b)

Fig. 5. TMM device for studying an involute gearing (a) and its 3D- model (b)

4 Conclusions

3D modeling of laboratory equipment (TMM-47 and other devices) forces a student to learn more in the field of TMM, combining the study of involute gearing theory, theory of cutting process, gear design and etc., along with application of computer programming and implementation of their own experience and knowledge of previous courses. From the point of view of Professors lecturing TMM, the proposed educational methodology gives more opportunities to a student and enlarges teaching by elements of augmented and virtual reality. The presented methodology is related to the information-and-developing technologies and aimed at the best comprehensibility of discipline, popularization and dissemination of scientific knowledge with an increasing effect in case of open access to 3D-documents hosted on Internet network.

References

1. Willis, R.: Principles of Mechanism. Longmans, Green, and Co., London (1870)
2. https://alchetron.com/Robert-Willis-(engineer)-1123803-W. Accessed 02 Sept 2017
3. LNCS Homepage. http://www.euler-2007.ch/doc/EulerLec.pdf. Accessed 02 Sept 2017
4. http://en.wikipedia.org/wiki/Involute_gear. Accessed 02 Sept 2017
5. Bogolyubov, N.N., Mikhailov, G.K., Yushkevich, A.P. (eds.): Euler and Modern Science, MAA Spectrum, Mathematical Association of America, Washington, D.C., translated from the Russian by Robert Burns (2007)
6. http://www.docdatabase.net/more-leonhardeuler-hislifethemanandhisworks-euler-2007-1194008.html. Accessed 02 Sept 2017
7. LNCS Homepage. http://icm.mcs.kent.edu. Accessed 02 Sept 2017
8. https://wn.com/involute_gear. Accessed 02 Sept 2017
9. Timofeev G., Egorova O., Samoilova M.: Harmonic Drive. Evolution of Wave Gear Design. http://www.iftomm2015.tw/IFToMM2015CD/PDF/OS7-001.pdf
10. Gautschi, W.: Leonhard Euler: his life, the man, and his works. SIAM Rev. **50**(1), 3–33. Published electronically
11. https://khkgears.net/new/gear_knowledge/gear-nomenclature.html. Accessed 06 Jan 2018
12. www.qtcgears.com/tools/catalogs/PDF_Q420/Tech.pdf. Accessed 06 Jan 2018
13. Romanov, V.F.: Calculations of metal-cutting tools. Mechanical engineering, Moscow (1969)
14. Theory of Mechanisms and Mechanics of Machines. Course of design: textbook under the editorship of Timofeev G.A., BMSTU, Moscow (2012). (in Russian)
15. Egorova, O., Omelchenko V., Kuzchbah A.: An Efficient Educational Methodology for Teaching "Theory of Machines and Mechanisms" Discipline, ISBN for the Proceedings: 978-986-04-6098/PDF/OS5-008, 2015, IFToMM 14th World Congress, Taipei, Taiwan (2015)
16. Egorova, O.: Three-dimensional computer modeling in teaching the «theory of mechanisms and machines» discipline. J. Izvestie VUZov Mech. Eng. **7**, 48–55 (2015)
17. Evgrafov, A., Khisamov, A., Egorova, O.: Experience of modernization of the curriculum TMM in St. Petersburg State Polytechnical University. In: New Trends in Educational Activity in the Field of Mechanism and Machine Theory Springer, ISEMMS-2013, pp. 239–247 (2014)

18. Egorova, O., Evgrafov, A., Rayushkina, V.: modern view on "history of science" teaching at Technical Universities. In: New Trends in Educational Activity in the Field of Mechanism and Machine Theory Springer, ISEMMS-2013, pp. 137–144 (2014)
19. Egorova, O., Vukolov, A.: New challenges of 3D mechanisms models in TMM training. In: International Review of Mechanical Engineering (IREME), Special Issue «The role of History of MMS in evolution of Mechanics», vol. 10, no. 5, pp. 303–311 (2016)
20. https://www.europeana.eu/portal/lv/record/2020801/dmglib_handler_docum_2240009. Accessed 06 Jan 2018

Cultural and Educational Significance of MMS Competitions for Future Engineers

V. Balyakin[1(✉)] and E. Krylov[2]

[1] Samara State Aerospace University, Samara, Russia
029-029@mail.ru
[2] Kalashnikov Izhevsk State Technical University, Izhevsk, Russia
649526@mail.ru

Abstract. The paper discusses the competencies of future mechanical engineer, and shows how the MMS student competitions promote the development of engineering knowledge and skills. Aspects of gained competencies are illustrated with problems and examples.

Keywords: Engineering · Education · MMS · IFToMM · Student competition Personal qualities of engineer

These days whatever Electronics, Informatics and other related branches of Science will be enhanced and expanded in Engineering and Technology, Machine and Mechanism Science (MMS) will be a foundation for design and manufacture of goods and objects human being needs of. This idea is clear for engineers and academics. But often it is not so evident for junior or sophomore students because of enhance of electronic technologies and devices in everyday life of members of modern society.

How to start and help the motivation towards MMS learning, especially if we think about the brightest and smartest students keen in different sciences? Taking apart the economic aspects such a competitive future salary (responsibility of society as a whole, rather than of academic frames as its part), one can appeal to the significance of MMS learning for personality, affecting his/her knowledge, skills, abilities, ways of thinking and other mental resources.

As we hope, during study years the expectations of the best mechanical engineering students are to gain qualities of ***analyst*** (analysis of past ideas and experience, discover open problems and challenges), ***researcher*** (understanding of operation, performance, economics, interaction of systems), ***designer*** (creation of alternative and innovative solutions within given limitations and constraints), ***computing engineer*** (simulation, estimation of systems, and solving the models using most productive software environment for engineers and scientists), ***effective communicator*** (both oral and written, in native and foreign language), ***team-worker*** (collaborative way of professional activity), ***open-minded person*** (ability of thinking outside national frame, long-life education), and even more.

Yet, these qualities and abilities are of demand by Industry, Research and Economic institutions, so they are enlisted in national Professional Standards of Engineer in different countries.

J. C. García-Prada and C. Castejón (Eds.): *New Trends in Educational Activity in the Field of Mechanism and Machine Theory*, MMS 64, pp. 38–48, 2019.
https://doi.org/10.1007/978-3-030-00108-7_5

Every mechanical discipline of bachelor curriculum contributes to making competencies relevant to listed above personal qualities. Of course, the most effective way should include some integration activity during senior courses. Unfortunately, the tendency exists (at least in Russian high education system) of reduction of teaching hours within the syllabi of mechanical engineering fundamentals. For example the MMS contact hours were reduced by **more than half** during last 10 years in Russian technical universities, mainly for practical lessons and labs. Lectures along are not successive to provide students with knowledge, ready to use in practical situation and during making MMS design project. The result is that a person, planning to make effective research and design project in Master program, often has not cultural and educational level to cope with it.

So, new approaches are required to provide students with necessary integrative knowledge's and abilities. In the leading Russian technical universities, the method of teaching students wishing to obtain high-quality knowledge in the field of the MMS, through the preparation for the Olympiads of various levels, has become widespread. The initial stage of the MMS Olympiad all interested students take part in the competition. Prior to this tour, there are consultations of leading teachers for students. After the first stage of the Olympiad, where the standard tasks of advanced complexity are solved, the examination of the participant solutions allows to identify the most successful students. From their number, those who wish to study the theory of mechanisms and machines are selected, and teams are formed to participate in the second stage of the Olympiad. At the **regional stage** of the Olympiad, more complex tasks are offered on the main topics studied in the theoretical course. When preparing for this stage of the Olympiad, the group of student has additional classes with the tutor. After the regional round of the Olympiad, the team of the most gifted students is selected and preparations are made for the **all-Russian tour** of the Olympiad. In the further preparation of the team, all leading teachers of the University take part. For each topic offered at the Russian Olympics round, classes are held by the teacher, who has the highest competence in this topic. As a rule, this is related to the topic of the research work of the teacher. For the independent preparation of students it is necessary to have a set of tasks for all the topics under consideration [1, 4]. These tasks on the one hand should concern all the main theoretical topics, and on the other hand should be original and interesting for students.

Following the fundamental mission of IFToMM "to promote research, development, and education in the field of Machines and Mechanisms using theoretical and experimental methods, along with their practical application", IFToMM Executive Council in Guanajuato, Mexico in June 2009 had established a regular **international** public MMS competition among students – Student International Olympiad on MMS (SIOMMS). The three first SIOMMS were successfully held in Russia (Izhevsk, 2011) [2, 3], China (Shanghai, 2013), and Spain (Madrid, 2016).

During the preparation period for regional, all-Russian, or international Olympiad students have additional classes in small groups. Actually time of learning MMS is doubled, compared with a standard curriculum. The intensity of thinking, brainstorming, complex research questions contained in the problems to be solved, - all these aspects contribute to intellectual facilities of future engineers.

What is important about the preparation for MMS competitions, that students involved are of second and third year, so they get a chance to participate in actual investigations – not of such high level, as scientifical research for Master or PhD degree, yet investigations of due level at due time of study life. And, rephrase the sentence, *winning isn't everything, it's the taking part that counts,* any participant wins because he/she changes himself/herself in the process of preparation. Of course, winning the competition is the super result.

Some of simple strategies leading towards desired qualities and abilities of participants are illustrated below by results, obtained by students themselves during preparation period.

- ***Analyst***

To master all competition topics, student should gain information from different sources: lectures, textbooks, papers and make analysis on the base of actualizing intellectual operations of data analysis and comparison, classification, categorization, generalization, abstraction, making solution, estimation of results for correctness, and so on.

It happens that in kinematical analysis of linkages the method of instant center is not so popular; the Aronhold-Kennedy theorem is not mentioned in Russian textbooks. It was suggested to two students to discover the method and make one step forward – apply it for 5-bar linkage. The students' finding concerning instant center of velocities in kinematics of coplanar linkages are given below.

"The displacement of any point of the body performing general plane motion (GPM) can be represented as the geometric sum of translation along with a reference point and rotation around this point. Also, for every GPM body, there is a point called the instant center of velocity, whose velocity at a given moment is zero. By Euler's theorem, the velocities of body points are the same as if the body were rotating around an axis passing through the instant velocity center perpendicular to the plane of motion. The method is well applied for the velocity analysis of a mechanism since any two bodies moving relative to one another have an instant center.

In coplanar mechanisms *primary* instant centers (permanent or fixed) can be identified simply by inspection of the kinematic pairs, for example pin connections, and rolling bodies with no slipping. *Secondary* instant centers (changing positions while the mechanism moves) are found using Kennedy's theorem [6].

Consider five-bar linkage (Fig. 1a) with known sizes and given angular velocities of two links, ω_2 = 10 rad/s и ω_5 = 5 rad/s. The problem is to determine velocity of pin C at the position shown.

The number of instant centers is determined from the formula N = [n(n − 1)]/2, where n—number of elements of the system. We have n = 5, hence N = 10.

In the Kennedy's circle (Fig. 1b) pins are located on the perimeter, and all possible pairs of elements are connected by straight lines. For the linkage given, there are five primary instant centers, I_{12}, I_{23}, I_{34}, I_{45}, I_{15}.

The secondary instant centers are found using the Aronhold-Kennedy theorem named after Siegfried Aronhold who proved it in 1872 and Alexander Kennedy, who proved it in 1886. The theorem states that the three instant centers shared by three rigid bodies in relative motion to one another all lie on the same straight line [5]. Following

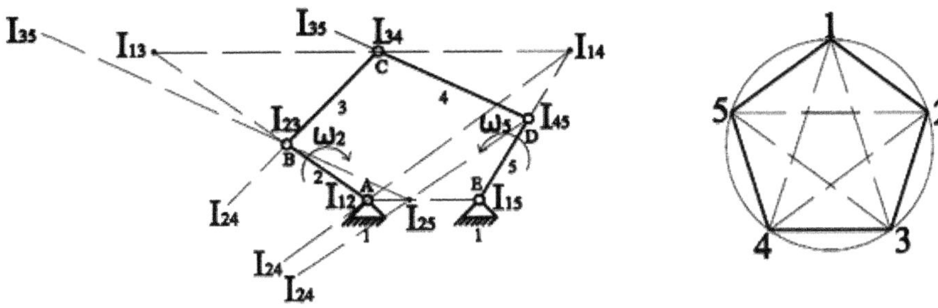

a *b*

Fig. 1. Locating instant centers of five-bar linkage

the theorem we should find the secondary center I_{25} somewhere on the straight line passing through points I_{12} and I_{15}.

According to the theorem about angular velocity relation [6], the ratio of ω_2 and ω_5 is negative, hence center I_{25} is located between I_{12} and I_{15}:

$$\frac{\omega_2}{\omega_5} = \frac{|I_{25}I_{15}|}{|I_{25}I_{12}|} = -2, \quad |I_{25}I_{15}| = -2 \cdot |I_{25}I_{12}|.$$

If $|I_{15}I_{12}| = 9$ cm, then center I_{25} is located at $|I_{25}I_{15}| = 6$ cm from center I_{15}.

The remaining instant centers can be found directly from the Aronhold-Kennedy theorem. The secondary center I_{13}, for example, must lie on the straight line $I_{23}I_{12}$. In addition, the secondary center I_{14} should lie on the straight line passing through points I_{34} and I_{14}. The location of all the secondary centers is shown in Fig. 2b.

To determine the velocity of the pin C, we need the angular velocity ω_3:$\omega_3 = \frac{\upsilon_B}{|I_{13}I_{23}|}$. Velocity of pin B is found as: $\upsilon_B = \omega_2 \cdot |I_{12}I_{23}|$.

Then the desired velocity of pin C equals:

$$\upsilon_C = \omega_2 \cdot \frac{|I_{12}I_{23}| \cdot |I_{13}I_{34}|}{|I_{13}I_{23}|}.$$

The students had revealed that the concept of an instant velocity center can be (having practiced a little) used by students in determining the angular velocities of links in multi-link mechanisms. This method allows one to make the solution of a kinematic problem shorter, and analyze the motion of links relative to one another."

- ***Researcher***

Engineering is a field devoted to understanding how things work. MMS is, in some sense, the successor of theoretical mechanics (TM = Statics, Kinematics, and Dynamics) in the curriculum, especially for methods of kinematical and dynamical analysis. But MMS studies actually existing objects, while TM is free to operate with rather simplified models. So, future researcher should practices in investigation of systems and making reasonable models.

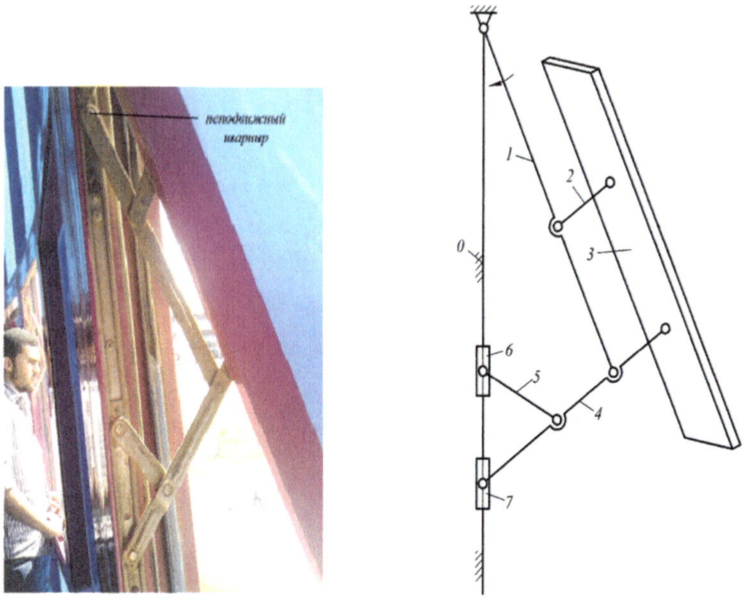

a *b*

Fig. 2. Actual mechanism and its possible structural scheme

A simple exercise to investigate a structure is shown below [4].

Problem: the photo (Fig. 2a) shows the mechanism of opening and closing the window frame. It is required: draw a structural diagram of the mechanism; determine the mechanism mobility; find the Assur groups, determine class of every group; write down the structural of the mechanism.

- ***Designer***

Design engineer researches and develops ideas and systems for manufacture, works to improve the performance and efficiency of existing products, research new developments and innovations. We consider here only one example, whereas one can find a lot of design problems in any MMS topic.

Problem: For a four-bar linkage the velocity diagram has been constructed that looks like an equilateral triangle, Fig. 3 [4]. It is known that the length of the crank *OA*, connecting rod *AB* and rocker arm *BC* are in the ratio $l_{AB} = l_{BC} = 3l_{OA}$. Velocity of crank point *A* is $V_A = 1$ m/s, angular velocity of the crank is $\omega_1 = 20$ rad/s.

It is required:

1. Design all possible variants of the mechanism scheme matching the velocity diagram.
2. Find the length of the fixed member *OC*.
3. For one of the variants determine the angular velocity of the rocker.
4. Identify crank-rocker and two-rocker mechanisms.

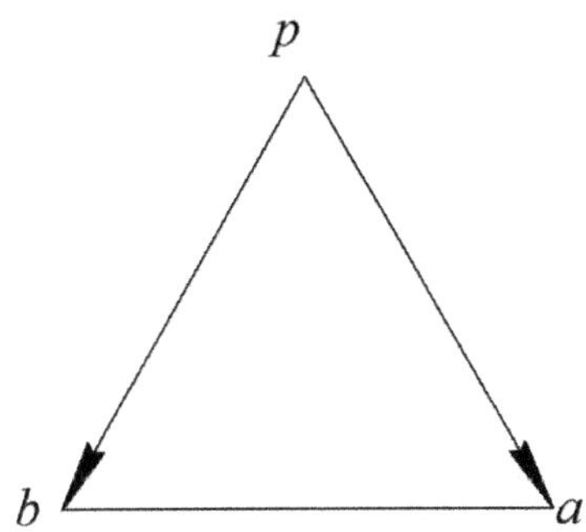

Fig. 3. Velocity diagram

The *solution* tells that four mechanism schemes corresponds to the specified velocity diagram, which are shown to scale in Fig. 4.

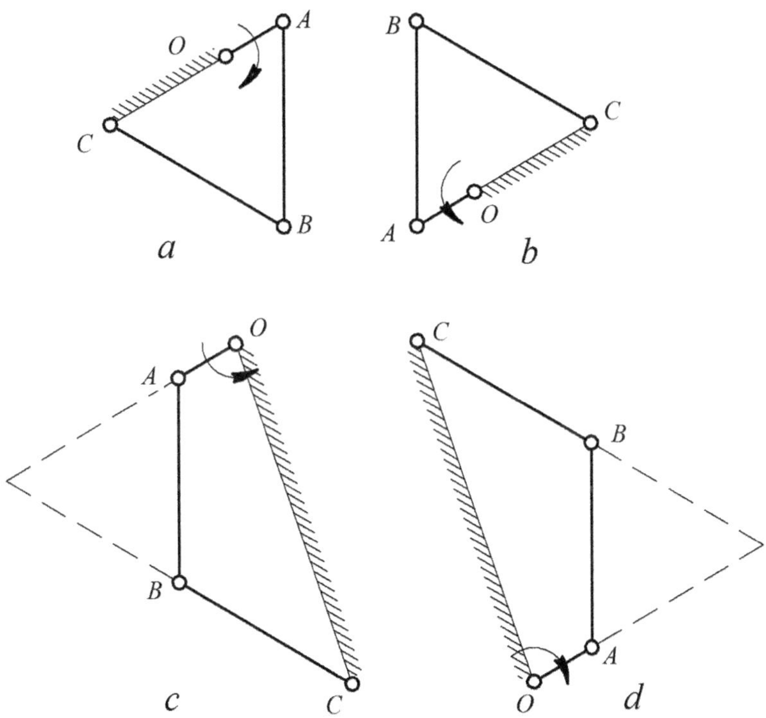

Fig. 4. Four-bar mechanism schemes

The schemes in Fig. 5 illustrate crank-rocker (*a*) and two-rocker (*b*) mechanisms.

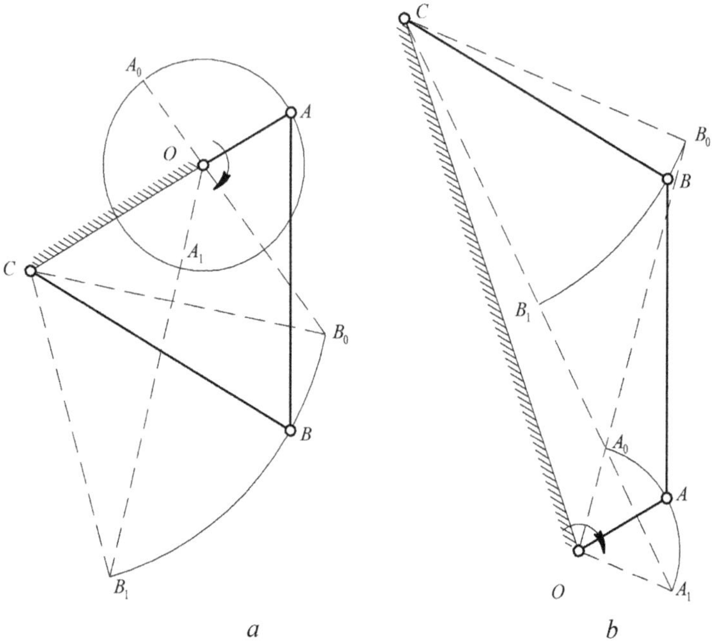

Fig. 5. Crank-rocker (a) and two-rocker (b) mechanisms

Discussions of MMS design problems provides the chances for students to gain the ability of recognizing facts of design in every-day life, modify and test designs, and turn the research ideas into technical plans.

- ***Computing engineer***

Mechanical engineer is expected not only to make a system simulation with the set of algebraic/vector/differential equations, the simulation should be completed with a numerical solution. That means that engineer should be keen with all aspects of the numerical solution of a problem, including numerical codes for available software systems. Sometimes that's a problem for students.

Preparation for MMS competition provides students with rich possibilities to practice with numerical solutions. We consider here only example in the field of dimensional synthesis of linkages.

Problem: make a graphical dimensional synthesis of a four-bar linkage for three given positions of crank *OA* (Fig. 6) and three corresponding positions of rocker *CD*. It is given that $A_1D_1 \neq A_2D_2 \neq A_3D_3$ [4].

Since $A_1D_1 \neq A_2D_2 \neq A_3D_3$, then the desired linkage should be of type shown in Fig. 7. The typical graphical solution [4] is based the method of inversion of motion, stopping link *CD* at the first position, and finding where point *A* is located relative to *CD* at the mechanism's second and third positions.

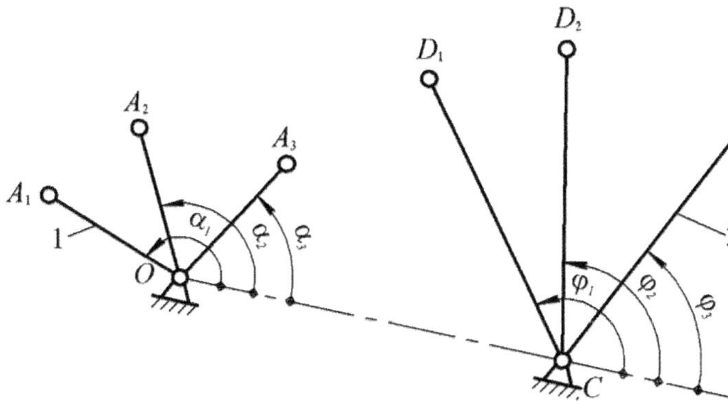

Fig. 6. Problem for dimensional synthesis of four-bar linkage

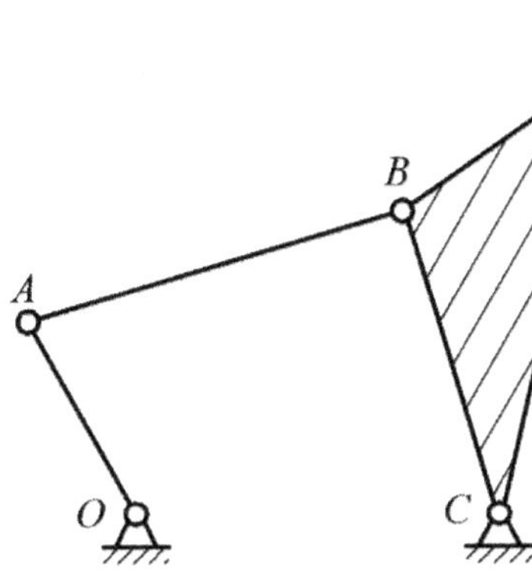

Fig. 7. Assumed linkage

The future participants of MMS competitions were asked to make numerical codes of the problem in Mathcad software to generalize the solution a somewhat. The solution obtained by the students is briefly listed below.

In the inversed motion, the rocker CD_1 is fixed, that is the plane containing points A_i (i = 2, 3) rotates about C for angles ($\varphi_1 - \varphi_i$). Then the new position vectors of A are determined by (the vector values are in **bold**):

$$\boldsymbol{OA}_i = Rot(\varphi_1 - \varphi_i)(\boldsymbol{OA}_i - \boldsymbol{OC}) + \boldsymbol{OC} \quad (i = 2, 3),$$

where Rot ($\varphi_1 - \varphi_i$) is the rotation transformation matrix.

Pin B_1 is the center of a circle of radius R = AB, described by the equations

$$(\boldsymbol{OA}_i - \boldsymbol{OB}_1)^2 = R^2, \quad i = 1, 2, 3.$$

Subtracting the expressions corresponding to $i = 2, 3$ from one with $i = 1$, we get

$$2(\boldsymbol{OA_2} - \boldsymbol{OA_1})\,\boldsymbol{OB_1} = (OA_2)^2 - (OA_1)^2,$$
$$2(\boldsymbol{OA_3} - \boldsymbol{OA_1})\,\boldsymbol{OB_1} = (OA_3)^2 - (OA_1)^2.$$

With introducing the matrices

$$M = 2\begin{pmatrix} (OA_2 - OA_1)_x & (OA_2 - OA_1)_y \\ (OA_3 - OA_1)_x & (OA_3 - OA_1)_y \end{pmatrix}, \quad V = \begin{pmatrix} OA_2^2 - OA_1^2 \\ OA_3^2 - OA_1^2 \end{pmatrix},$$

the system is transformed into

$$M \cdot \boldsymbol{OB_1} = \boldsymbol{V}, \quad \text{or} \quad \boldsymbol{OB_1} = M^{-1} \cdot \boldsymbol{V}.$$

Positions of points B_i (i = 2, 3) are found by rotating the vector passing through B_1, through angle $(\varphi_i - \varphi_1)$ about fixed point C. That is,

$$\boldsymbol{OB}_i = \mathrm{Rot}(\varphi_i - \varphi_1)(M^{-1} \cdot \boldsymbol{V} - \boldsymbol{OC}) + \boldsymbol{OC}\ (i = 2, 3). \quad \textbf{Answer}$$

To visualize the set of successive positions of the mechanism, i.e. to make the animation, consider the loop contour $OABC$:

$$(\boldsymbol{OC} - \boldsymbol{BC} - \boldsymbol{OA})^2 = \boldsymbol{AB}^2 = \text{const}.$$

Expanding the expression, we get

$$OC^2 + BC^2 + OA^2 + 2OA \cdot BC \cdot cos(\theta - \alpha) - 2OC \cdot BC \cdot cos\theta - 2OC \cdot OA \cdot cos\alpha$$
$$= AB^2.$$

Here θ specifies angular position of **CB** with respect to **OC**.
Differentiating by time, we have

$$OA \cdot BC \cdot sin(\theta - \alpha)(\dot{\theta} - \dot{\alpha}) - OC \cdot BC \cdot sin\theta \cdot \dot{\theta} - OC \cdot OA \cdot sin\alpha \cdot \dot{\alpha} = 0.$$

The differential equation $\frac{d\theta}{d\alpha} = \mathrm{F}(\alpha, \theta)$ is solved by the Runge–Kutta method for the array of θ_i as a function of α_i. The graphical image of synthesized mechanism is shown in Fig. 8a.

Solving the problem numerically, the students understood the method and applied it to quickly solve other problems. So, Fig. 8b illustrates the animation of off-set crank and slider mechanism, as suggested in the problem set of SIOMMS-2013 (Shanghai, China).

- ***Effective communicator***

The participants of the MMS competitions should present their solutions to the jury board as readable texts supplied by due pictures and formulas. In other words, the

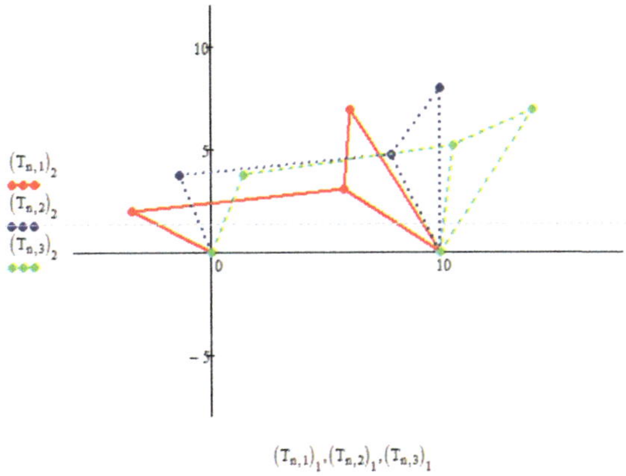

a

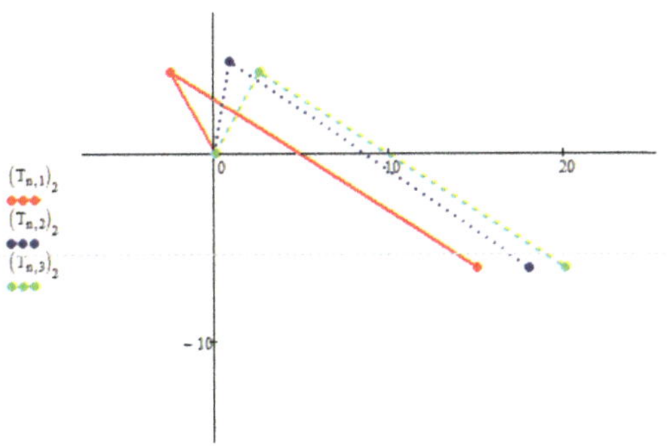

b

Fig. 8. Animation of the synthesized mechanisms

students should be ready to communicate the solutions and to manage the exchange of information and ideas with other peoples. That's not quite simple for some persons, especially in the situations of foreign-language communication.

So, the preparation period is important in communicative aspects not less than in science ones. It happens nowadays that English is a *franqua ligua* for modern science, and to keep pace with recent engineering topics, one should learn English. Unfortunately, English skills of average Russian student are far from excellent. One of Russian participants of SIOMMS-2016 (Madrid, Spain) was so tired with intensive English during the event, that failed to analyze texts of two contest problems ("*I did not understand what they wanted from me at all*"). There is a need to make language training more effective not to lose positions in the international competitions.

English communication in the preparation period has a profound influence on the personality of the students in communicative sense.

- ***Team-worker***

Another benefit from preparation for MMS competitions is cooperation of students in small (3–6 persons) groups. Often the contest problems contain several aspects, that makes a room for splitting the elements of solution between students. Subsequent discussions and synthesis of the solution contribute to collaborative behavior of them. For a tutor it is a great pleasure to discuss the topics with group of smart young persons.

- ***Open-minded person***

Effective engineer should be open to the world of science, technology, recent facts, knowledge, ets., not limited by national frontiers. The name and works of Franz Reuleaux should be known to Russian students not less than the name and works of Nikolay Zhukovsky to German students. Eventually, Chebyshev's structural formula, universal Somov-Malushev's mobility equation, and Grübler–Kutzbach's criterion are about the same issue of one of MMS topics.

1 Conclusion

The competencies, obtained by students during the preparation and participation in MMS competitions, are relevant to qualities and abilities demanded by industry, science, and economy. The competitions promote MMS among all body of students; provide the possibilities for the smartest young persons to apply interdisciplinary skills and make a synthesis of gained knowledge; make a deep positive cultural and educational effect on future engineer.

References

1. Artobolevsky, I.I.: Theory of Mechanisms and Machines, 4th edn. Science (2009). 639 p
2. Goldfarb, V.I, Krylov, E.G., Elensky, A.V.: The first student international olympiad on mechanism and machine science – the challenge in MMS education. In: The 2nd IFToMM Asian Conference on Mechanism and Machine Science, 7–10 Nov 2012, Tokyo, Japan
3. Goldfarb, V.I., Krylov, E.G., Elenskii, A.V.: Analysis of the participant solutions of the first student international Olympiad on mechanism and machine science. Mech. Mach. Theory **70**, 293–297 (2013)
4. Kuznetsov, N.P. (ed.): Student Olympiads on Theory of Machines and Mechanisms. Problems and Solutions. Publishing House of Kalashnikov ISTU, Izhevsk (2013). 288 p
5. Khurmi, R.S., Gupta, J.K.: Theory of Machines, 14th edn. S Chand & Co Ltd., Chennai (2005). 1071 p
6. Uicker, J., Pennock, G., Shigley, J.: Theory of Machines and Mechanisms, 4th edn. Oxford University Press, Oxford (2011). 900 p

Project-Based Learning Applied to Mechatronics Teaching

A. Noriega(✉), J. M. Sierra, J. L. Cortizo, M. J. Prieto, F. F. Linera, and J. A. Martín

University of Oviedo, Gijón, Spain
noriegaalvaro@uniovi.es

Abstract. This paper deals with the experience carried out in the University of Oviedo with students of the Master in Mechatronic Engineering. In this master, there is a reduced number of students (never over 20), and their nationalities, as well as their previous academic background, can be very different. This results in groups of students who are usually eager to learn new things and in an encouraging and enriching environment –also for lecturers. Since most of the courses included in the Master have a practical focus, this spirit was also used in the subject called Mechatronic Project. In this subject, the students –supervised by Electronic Engineering and Mechanical Engineering lecturers– must develop a planar or spatial manipulator. The manipulator must be designed, manufactured and mounted throughout the subject and, during the evaluation session, it must correctly follow pre-defined trajectories. All the different students' groups in the course are provided with commercial components such as servomotors, ball bearings, linear guides; they can also use 3-D printers in order to design the manipulator links or its frame. They also design and manufacture the printed circuit board including the control circuitry for the servomotors, liquid-crystal displays, keyboards, connectors, etc. The present paper provides a detailed description of all the tasks carried out by the students: design, manufacture, mounting, defense and tests. Economical cost of the experience described, as well as some problems to be solved in the future, will also be evaluated. The final conclusion evinces that this experience has been very welcome both by the students and the lecturers involved, and that it has proved successful.

1 Introduction

The Master in Mechatronic Engineering (MME) is offered by the University of Oviedo and is the local support to the Erasmus Mundus Master in Mechatronics and Micromechatronic Systems (EU4M). Both masters have the same objectives: preparing students in theoretical and practical courses focused on mechanics, electronics, informatics and control so that they are able to design a mechatronic device, covering all the stages in the process, i.e. the conceptual design, the manufacturing process, the mounting process, and the test.

Both MME and EU4M are two-year master's degrees (four semesters with 120 ECTS). EU4M is taught in three European countries (Germany, France and Spain). In each of these countries, the studies are conducted in the local language, and the

J. C. García-Prada and C. Castejón (Eds.): *New Trends in Educational Activity in the Field of Mechanism and Machine Theory*, MMS 64, pp. 49–56, 2019.
https://doi.org/10.1007/978-3-030-00108-7_6

students of EU4M should stay in at least two of the three countries where it is offered. This helps students to learn more about the different countries and their local culture. All of the students in MME are Spanish whereas the students in EU4M mainly come from Asia and from Latin America.

The profile of the students is, in most of the cases, mechanical or electronic engineering, but there is a wide variety of profiles that can join the master: computer engineering students, graduates in physics, or even graduates in mechatronics. This diversity can be a problem because the initial knowledge each student has is different, but the experience of these years has shown that it is definitely a great advantage when they have to deal with a design project in work-groups.

The master was launched in year 2006–07, but its contents were annually adapted in meetings held by the EU4M Consortium until year 2011–12, when a general revision took place and defined a common base for the master in the three countries. It was precisely in year 2011–12 that the subject "Mechatronic Project" was first introduced in the programme of the first year, second semester. The subject has 5 ECTS credits assigned, and the staff (three lecturers) belongs to two different knowledge areas (mechanical engineering and electronic technology), which means that an additional effort is needed by the lecturers so that the final goals of the course are achieved [1]. The skills covered by this subject are the following:

Basic skills

- Students should be able to use their knowledge and ability to solve real problems in a new (or not-well-known) multidisciplinary context related to mechatronics.
- Students should be able to integrate knowledge and assume decisions from information that, being incomplete or limited, might include reflections about the social and ethic responsibility.
- Students should be able to communicate the conclusions of their works, based on technical knowledge, to specialized and non-specialized public, in a clear way.
- Students should be able to study in an autonomous way, and find new knowledge, or techniques that allow them to solve new problems in the future.

General skills

- Writing technical specifications to describe a mechatronic problem, from a global and inclusive point of view, via prescriptive specifications, or performance of procedures as appropriate.
- Organizing conceptual design activities, to help in the decision about the best solution.
- Working as design managers in mechatronic systems design, assuming functions of planning (setting goals and strategies), organization (assigning tasks and resources), direction (leading and motivating), and control (tracing).
- Planning and organizing a research activity in mechatronics.
- Improving skills to work in a collaborative and multidisciplinary group.

Specific skills

- Selecting materials, calculating and optimizing the dimensions of mechatronic systems according to stress and strength criteria, durability and reliability.

- Planning the manufacturing process of mechatronic systems, according to the specifications, work environment, availability, and mounting and verification processes.
- Selecting the programmable logic devices needed in each case, to design mechatronic systems, managing the development process and designing the interface stages needed.

Learning results:

- Students should be able to propose and develop an appropriate mechanism in the machine design process, to solve a problem, taking into account the relationship with other parts of the machine and the control implications.
- Students should be able to develop a virtual prototype of the mechatronic system, taking into account physical requirements, control elements and system feedback that allows the goodness of the design to be verified.
- Students will be expert users of databases for mechanic, electric or electronic components used in mechatronic systems design.
- Students will be able to write technical specifications in order to document the design work done and technically support the decisions assumed.
- Students will be able to search international databases and find information of new inventions, technical journals, or other technical information that can be used in a project.

As it can be seen in the skills described, this subject aims to improve the higher order thinking skills defined in Bloom's taxonomy shown in Table 1, in its original [2] and revised [3] version. According to the revised version, the capability to "create" implies: design, manufacture, plan and drawing.

Table 1. Original and revised Bloom's taxonomy

Higher order thinking skills	*Evaluation*	*Create*
⇧	*Synthesis*	*Evaluate*
	Analysis	*Analyze*
	Application	*Apply*
	Understanding	*Understand*
Lower order thinking skills	*Knowledge*	*Remember*
	Bloom's taxonomy	**Revised Bloom's taxonomy**

2 Methodology

The topics to be dealt with in the subject have been developed around a project in order to apply the so-called "project-based learning" methodology [4]. The topics have a theoretical part and associated practical sessions that are oriented to the project of the course. Figure 1 shows the topics of the course, the part of the design project involved, the type of session (theoretical, practice, tutorial) and the knowledge area involved.

TOPICS IN THE OFICIAL TEACHING GUIDE

PROJECT STAGES	1. Technical specifications and project documentation			2. Sketching, modelization, simulation and drawings in a project			3. Workgroup, management, planning and monitoring of the project			4. Innovation tools based on databases, journals, webs,...			5. Protection of results, patents and utility models			6. Development of a mechatronic prototype			7. Documentation and public presentation of project results		
	CE	PL	TG	CE	PL	TG	CE	PL	TG	CE	PL	TG	CE	PL	TG	CE	PL	TG	CE	PL	TG
1. Technical specifications	1	1																			
2. Planning and task organization																					
2.1 Gantt and Pert							1	1													
3. Design																					
3.1 Design concepts				1	1																
3.2 Synthesis and analysis				1	1	1												2			
3.3 Modelization and simulation				1														1			
3.4 Selection of mechanical components										1	1					1					
3.5 Communication by I2C bus																1		3			
3.6 Use of servomotors																1		2			
3.7 Limit switches																1					
4. Drawings, bills of materials and calculations annexes																					
4.1 Mechanical	1	1																			
4.2 Electronic				1	1	1															
5. Purchase, manufacture and assembly of components																		2			
6. Official documentation			1																1		
7. Protection of results													1								
8. Public presentation																					1
	CE	PL	TG	CE	PL	TG	CE	PL	TG	CE	PL	TG	CE	PL	TG	CE	PL	TG	CE	PL	TG
TOTAL IN TEACHING GUIDE:	*2*	*2*	*1*	*4*	*3*	*2*	*1*	*1*	*0*	*1*	*1*	*0*	*1*	*0*	*0*	*4*	*0*	*10*	*1*	*0*	*1*
REAL DISTRIBUTION OF SESSIONS:	**2**	**2**	**1**	**4**	**3**	**2**	**1**	**1**	**0**	**1**	**1**	**0**	**1**	**0**	**0**	**4**	**0**	**10**	**1**	**0**	**1**

Mechanical Engineering Area

Electronic Tecnology Area

CE: Theoretical class
PL: Laboratory class
TG: Group tutoring

Fig. 1. Topics, types of classes and distribution of lecturers.

The project focuses on the design and manufacturing of a manipulator with 2 or 3 degrees of freedom that must be able to describe several predefined trajectories with the tool center point. The students can select the type of manipulator from several possibilities. Every year, the desired trajectories and the allowed types of manipulator are changed.

To carry out the project, the students need to apply the knowledge acquired in some of the subjects of the first semester of the master in order to design and manufacture the manipulator. Finally, they have to prepare the technical documentation of the project, manufacture the prototype and show the work done in a public presentation (including a functional test of the prototype designed) to a tribunal formed by the three lecturers of the subject.

The students have to work in groups of two, three or four people (depending on the year). Every group must include at least one student from each master (MME and EU4M) in order to facilitate the cultural integration. Also, every group must include at least one student with a background in mechanics and other one with a background in electronics in order to increase the power of the team with a wide range of knowledge in different disciplines [5]; this also contributes to the experience exchange between students. Under these conditions, three or four different work-groups are made every year.

The groups are responsible for preparing their own work plan and dividing the work between the group members. This plan should meet some temporary requirements imposed by the lecturers, such as time period to use the manufacture facilities or date to finalize the design, for instance. These temporary limitations are imposed in order to make students aware that, in the real world, time is limited and unexpected problems might affect the final result.

All the groups do their planning and present it at the beginning of the project. Capability of each student to perform their tasks in the time period assigned ac-counts for part of their final qualification. The other part depends on the final prototype developed by the group.

Throughout the course, the students are formed in different design techniques, some of them being absolutely necessary for the final design to be carried out (for instance, controlling servomotors) while others are only useful tools the students could decide whether to use or not. In tutorial periods the lecturers revise the work done and help students in order to find the optimal technical and economical solution. Also, they encourage the students to be critic with their design decisions.

To carry out the design and manufacturing of the manipulators, the students are provided with a set of hand tools and are allowed to use commercial components (such as servomotors, bearings, linear guides, screws and nuts). They can also use 3D printers, with Fusion Deposition Modeling (FDM) technology to manufacture the pieces with a complex geometry in Acrylonitrile Butadiene Styrene (ABS); thus, they do not need to weld or machine metal pieces. The pieces made with the 3D printers are light, low-cost and allow relatively complex geometries to be defined. The drawback is that printed pieces have not much toughness and accuracy, so students usually need to manually fit these pieces.

The final mounting and testing period usually takes a long time due to several reasons: mistakes or delays in the reception of commercial components, delays in the manufacturing of pieces, problems of integration of electronic components, …

To measure the learning of the students in this project, two surveys are run yearly. The initial survey is made the first day of class and includes questions about general information of the subject and the university, previous experience working in group, oral communication skill and self-learning. The final survey is conducted the last day in order to ask the students about the motivation during the project, the experience working in group, learning methodology, integration with other subject, available means and general assessment. The aim is to collect personal opinions and to compare the initial and final points of view in order to assess the real learning perceived by the students.

The surveys defined to evaluate the performance of the students were approved by the MME-EU4M Joint Academic Committee. This organism, formed by the coordinators of the *Master in Mechatronic Engineering, MME,* and the *Erasmus Mundus Master in Mechatronics and Micromechatronic Systems, EU4M,* together with five professors from these degrees and one Student Representative from each master degree, is in charge, among other tasks, of guaranteeing the ethics of all the surveys carried out in the aforementioned degrees. The Committee ensures that surveys meet three requirements: voluntary participation, no damage to participants (the surveys shall not affect the results obtained) and anonymity and confidentiality. All the three requirements were met by the surveys used to evaluate the results presented in this paper.

As far as the students are concerned, they were informed about all the survey details at the beginning of the course. Particularly, they were told that participation was not compulsory; thus, it was agreed by all parties that involvement of the students in the process would be an indicator that those students were aware of the conditions of the survey and accepted them.

3 Results

Throughout the four-year experience considered in this paper, there was only one group that did not get to manufacture and mount their manipulator. The rest of groups developed the project in the specified period and the results were successful, obtaining, in general, high qualifications. In Fig 2. some examples of manipulators built by the students are shown.

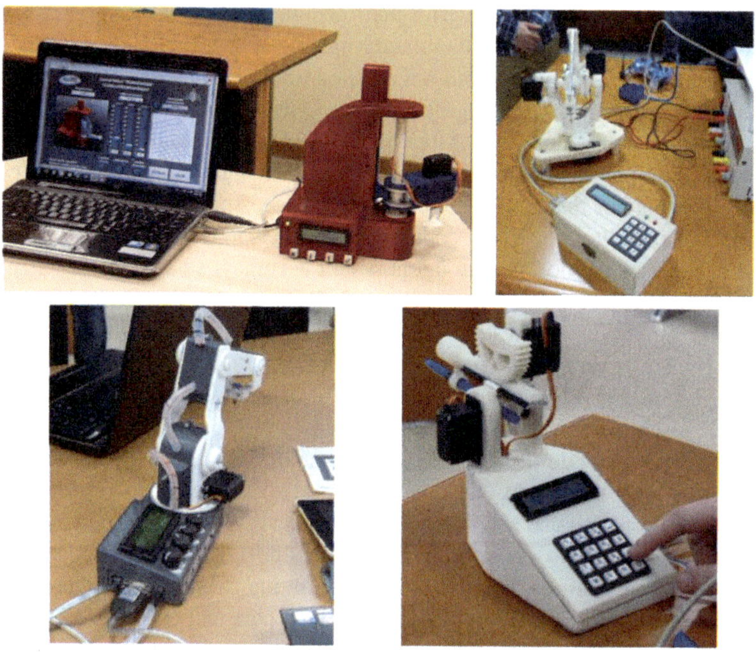

Fig. 2. Manipulators built by the students.

The results of the surveys allow the authors to extract the following conclusions:

- The competition between teams has been very motivating for a high percentage of students (around 65%), but, at the same time, there is a small percentage (around 15%) for which the competition has a negative effect.
- The topic of the project has had a significant influence on the student's motivation.
- Students have had great difficulties in establishing a hierarchical structure in a work-group formed by themselves.
- A very high percentage of the groups have not been able to execute the task planning. The main reason they alleged is the very optimistic forecast about manufacture and assembly times.
- A very significant percentage of students (around 36%) perceived they had improved their skills to make technical documentation and oral presentations compared to those skills obtained in their previous degrees.

- Students enjoyed working in the project, and learned in a way different from that used in other subjects, in which they only had to study for a final exam.
- Students enjoyed working in a real design (including manufacturing and mounting tasks), and they dedicated many more hours than initially planned. This latter point might be regarded as an inconvenience, since the students were left with less time to study the other subjects in the same semester.

4 Discussion and Conclusions

The experience has been successful and well evaluated by both students and lecturers. Lecturers of different knowledge areas working together did a good work, which shows that this kind of collaboration is possible, and necessary for this subject. The experience was useful for other lecturers involved in the master course who are now planning to modify their lectures for the next years in order to focus on the most interesting aspects of mechatronic projects. Thus, the lack of a mechatronic area in the University of Oviedo might be somehow compensated by including more integrated courses between mechanics, electronics and control programmes. This is, after all, the main goal of the MME-EU4M masters.

However, there are some problems to solve in the future. To start with, the usual problem with work-groups: how to coordinate the individual work of each student into the group to achieve the global goal of the project. New rules are needed in order to create a hierarchical structure in the group similar to the one existing in a real project of engineering.

Another problem is cost. The final cost of the prototypes was about 300–400 € each, depending on the year and the project so it can only be done in small groups of students. This might become a problem if the number of students increases in the future.

The students also said that, in order to fulfill their designs, they needed to learn more capabilities than the ones offered by the lecturers in this course; some of these capabilities might be added to the course in the future. This means that lecturers must make an effort to adapt their subjects to the new features required to design and assemble the manipulators year after year.

Finally, there was another unexpected consequence to other subjects in the same semester: students have dedicated many more hours than initially planned to this subject. Thus, they have had less time to study these other subjects.

Acknowledgements. This work has been supported by the Erasmus+ Programme of the European Union.

References

1. Brown, N.J., Brown, O.T.: Mechatronics "a graduate perspective". Mechatronics **12**, 159–167 (2002)
2. Bloom, B.S., Engelhart, M.D., Furst, E.J., Hill, W.H., Krathwohl, D.R.: Taxonomy of Educational Objectives: The Classification of Educational Goals. Longman, New York (1956)
3. Anderson, L.W., Krathwohl, D. (eds.): A Taxonomy for Learning, Teaching and Assessing: A Revision of Bloom's Taxonomy of Educational Objectives. Longman, New York (2001)
4. Wang, Y., Yu, Y., Wiedmann, H., Xie, N., Xie, C., Jiang, W., Feng, X.: Project based learning in mechatronics education in close collaboration with industrial: methodologies, examples and experiences. Mechatronics **22**, 862–869 (2012)
5. Soares, F.O., Sepúlveda, M.J., Monteiro, S., Lima, R.M., Dinis-Carvalho, J.: An integrated project of entrepreneurship and innovation in engineering education. Mechatronics **23**(8), 987–996 (2013)

Learning Machine Diagnostics Through Laboratory Experiments

M. Zamorano(✉), M. J. Gómez, C. Castejón, and J. C. García-Prada

University Carlos III of Madrid, Madrid, Spain
mazamora@ing.uc3m.es

Abstract. Hands-on-experiments are an important part of education in engineers. This work presents the development of laboratory practices for testing related to signal acquisition tasks that have been performed with students. The tests have been developed in the framework of a subject, where signal processing tools applied to machine diagnostics are taught as an important part of maintenance. Specifically, the type of signals that are used in the subject are time domain vibration signals.

With the inclusion of the tests, postgraduate students of engineering learn how to use a machine to obtain vibration signals and how to set the machine for that purpose. Also, they are able to diagnose rotating elements with different faults; such as bearings, and unbalanced shafts.

The laboratory practice gives the opportunity to tackle a job for themselves, which is one of the most important skills for the future engineer, as they gain experience.

Keywords: Hands-on-experiments · Education · Machine diagnostics Signal processing · Easy learning

1 Introduction

Laboratory experiments are very important for engineering students training, and they have even more importance if they do them by their own, because they acquire the required knowledge to perform the whole process. It is also considered very important by students, because this is an excellent way to emphasize and implement what they are learning during the course.

One of the most important issues for the engineer curricula is practice [1]. Nevertheless, the cost and time required to set up a physical laboratory and the test is the drawback [2]. Due to this issue, some virtual laboratories which use vibration data (as an example) from standard DataBases [3] or simulated data from formulas are being more and more common in universities. In this cases the skill of experimentality is lost, and students do not face with the behavior study of real mechanical systems and the difference with theoretical ones.

In the case of the current work, some guided tests developed for laboratory practices are detailed. The learning results obtained with this laboratory practice, which are public domain, are also included. The tests were performed in the framework of the subject '*Analysis and Diagnostics of Machinery*' that is coursed in the first year of the

J. C. García-Prada and C. Castejón (Eds.): *New Trends in Educational Activity in the Field of Mechanism and Machine Theory*, MMS 64, pp. 57–63, 2019.
https://doi.org/10.1007/978-3-030-00108-7_7

Master in Industrial Mechanics of the *Universidad Carlos III de Madrid*. These students were previously notified of the study, but this is just an improvement that we have tried in order to know if the students improve their knowledge and if they get better marks. If so, this will be implemented in the following years. The aim of the subject is that the students learn advanced techniques for mechanical vibration analysis, mainly oriented to machine fault diagnostics. The subject is equivalent to 4 European Credit Transfer and Accumulation System (ECTS) [4], 60% of which corresponds to weekly classes, and the remaining 40% to homework, including the guided tests.

The weekly classes are organized in theory and practices. The theory is presented in an ordinary classroom and the next week some practices about the theory are performed in a computer lab using Matlab®.

The tests are done by a group work, as part of the homework, where they must perform one of the available tests; bearings, unbalanced shafts, cracked shafts, or different couplings. They must set the machine and acquire the signals under different conditions of fault (including healthy condition) and speed. They are also required to process the signals, and to take considerations about the possibilities of diagnostics of each fault, and about the best conditions for diagnostics. The mark of the group work represents the 40% of the whole evaluation of the subject.

The tests are done by the students under the supervision of the teacher, and they are also provided a detailed written guide. The guide describes safety conditions for the machine; its main parts: motor, coupling, regulator, bearings, brackets bearings housing and shaft; the measurement chain: accelerometer, signal conditioner, data acquisition card and computer; the start-up of the machine; the steps and theory of data collection and the four different tests. Each test is explained with the most important theory to understand better the process and it is also explained how to do the test, the set and configuration of the machine, the main features of the elements and the signal acquisition process.

Before the guided lab tests were introduced, students were given signals and they were able to process them. Currently, they not only learn signal processing, but also signal acquisition techniques and considerations that must be taken.

The sensorized bench used to put theory into practice is the Machinery Fault Simulator (MFS) of SpectraQuest Inc. [5]. In this machine, controlled and calibrated faults elements can be installed, allowing the study of the vibration spectra of common faults.

In this paper both the machine and the acquisition process are described. Moreover, an example of one of the proposed practices is described, as well as its results.

2 Description of the Machinery Fault Simulator

The MFS is an excellent tool for learning Machinery Diagnostics, as it has several applications thanks to its versatility and easy to use.

MFS consists of a motor, which moves a sliding shaft set on bearings, which are placed in split brackets bearing housing, all of which are designed to be easily removed and replaced between various experiments (see Fig. 1). It also has a speed regulator, which allows to work with different conditions.

Fig. 1. Main parts of the MFS

MFS provides a basic setup for performing experiments and learning vibration signatures of different machine malfunctions. It also has some optional kits to investigate more advanced vibration phenomena or machinery fault, such as shaft alignment training, balance training, bearing faults or cracks in shafts [5].

The machine includes a set of four coupling types with different stiffness. With these components students can learn the effects of coupling stiffness on rotor dynamics and vibration signature of various defects and clarify the complexities of machinery shaft misalignment problems.

MFS also contains an eccentric rotor (see Fig. 2). It is used to learn the effects of rotor eccentricity on vibration spectra and develop techniques, to locate and correct the effects of eccentricity and to learn the effect of varying the mass moment of inertia on vibration amplitude among other things.

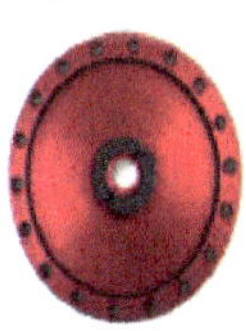

Fig. 2. Eccentric rotor

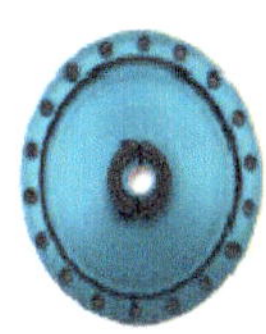

Fig. 3. Cocked rotor

A cocked rotor (see Fig. 3) can be used to learn the effects of a sheave that has not been fitted to the shaft properly, learn vibration signature of a cocked rotor or develop methods to correct cocked rotor problems.

With these and other rotors we can study unbalance, which is one of the main reasons of vibrations in machines.

In addition to these, it has numerous other kits that allow a more complete and specific study and the *training curriculum manual* [5].

3 Description of the Acquisition System

An acquisition system (see Fig. 4) is used to get the vibration signals in the computer, so they can be processed and compared with other signals.

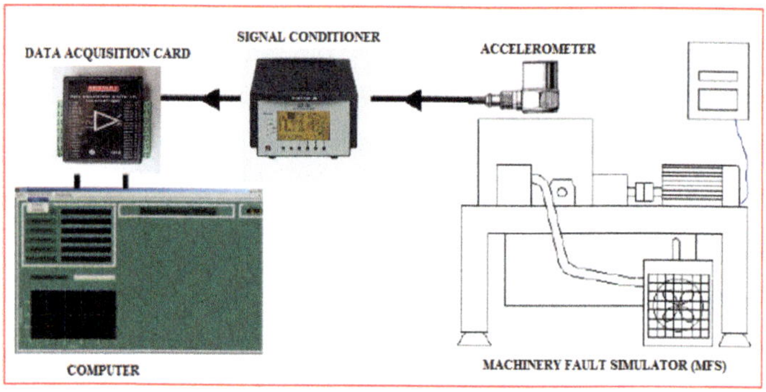

Fig. 4. Acquisition system

A piezoelectric accelerometer (Brüel&Kjaer 4383) is set on one of the brackets bearing housing. This is connected to a signal conditioner (Nexus Brüel&Kjaer 2693), which amplifies the signal coming from the accelerometer, and applies an antialiasing filter. The signal conditioner is connected to a data acquisition card (Keithley KUSB-3100), which transforms the analog signal into digital, so the computer can process it. The computer, therefore, is connected to that card via USB. The software used to obtain the signals is called Btool [6], developed in Matlab® environment. This software is easy and intuitive to be used by students and researchers.

4 Example and Results: Vibration Signals in Bearings

With the sensorized MFS, students can perform several tests related to defects in mechanical components. One of them is the simulation of new and faulty bearings [7].

The test consists in obtaining the vibration signals coming from bearings with different fault conditions. First, a healthy bearing is tested. Later, two bearings with inner-race fault and outer-race fault, respectively, are tested. The test consists in rotating each bearing with three different constant motor speeds: 20, 40 and 60 Hz.

As a part of learning, students must calculate theoretical defect frequencies BPFI (Ball Pass Frequency of the Inner race) and BPFO (Ball Pass frequency of the Outer race) with the appropriate formulas [8]. With this calculation they can focus on a specific range within the frequency spectrum of the vibration signal, which facilitates the diagnosis. Also, students can learn that theory does not always agree perfectly with reality. In Tables 1 and 2 we can see the theoretical BPFI and BPFO respectively, calculated with the formulas. We can also see the experimental frequencies, which are used to compare the amplitudes, because they are the closest frequencies to the theoretical ones with significant amplitude peaks.

Table 1. Theoretical and experimental frequencies BPFI

Defect frequencies (Hz)	Motor speed (Hz)		
	20	40	60
Theoretical (BPFI)	98.97	197.93	296.90
Experimental	99.98	195.2	293.4

Table 2. Theoretical and experimental frequencies BPFO

Defect frequencies (Hz)	Motor speed (Hz)		
	20	40	60
Theoretical (BPFO)	61.03	122.07	183.10
Experimental	60.43	121.6	182.8

To perform the tests, students place the accelerometer in the farthest bracket from the motor to avoid noises. Then, the acquisition system is connected setting the appropriate characteristics.

At each speed, 100 vibration signals are obtained. Power spectral density (PSD) of the Hilbert transform [9] of all the signals is processed and averaged by groups of 100 using an interface, specially developed in Matlab®.

As an example of the results obtained by students, the average PSD for the case of a bearing with outer-race defect at 60 Hz is shown in Fig. 5.

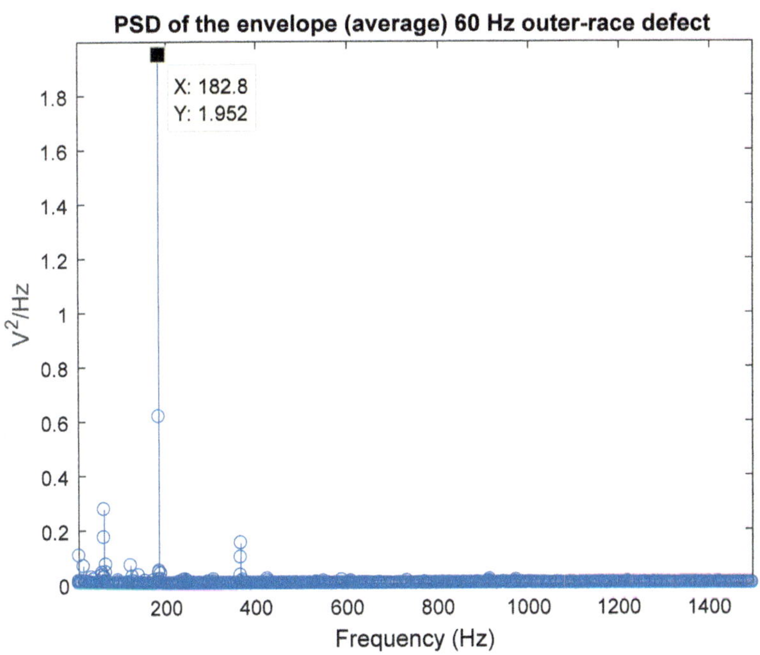

Fig. 5. PSD of the envelope (average) 60 Hz outer-race defect

In the figure, students can easily see a large amplitude peak near the theoretical BPFO, which is the experimental frequency shown in Table 2. Then, students concluded that the defective bearing tested has a defect in the outer race.

5 Students Learning Results

Three groups of students performed the tests. One group tested healthy and faulty bearings, the second unbalanced shafts and the third group cracked shafts [10, 11]. The experience showed that students were able to set the machine and the acquisition system using the guide provided without any problems.

Students presented a final report, including proper explanations about the signal acquisition process and the tests. Very good results of machine diagnostics were obtained. They showed that the knowledge acquired was really satisfactory. The teachers have detected that the quality of the reports has increased with respect to the previous courses. This was transmitted in the marks of the students, which have improved. With respect to the previous course, the average mark of the work group has improved a 23%. The final marks of the students have improved in an 18%.

The results of student surveys about the subject showed that their satisfaction level is higher than in previous courses. The average final mark of the subject was 12% higher than in the previous course, where the need of experimental lab tests was highlighted by the students.

6 Conclusions

In this work, the inclusion of laboratory practice for learning machine fault diagnostics is presented. The practice, carried out in the laboratory, let not only apply signal processing using a software, but also faces to the real problem of acquisition data from a mechanical system which is one of the lack of experience that students have. MFS is very useful for students due to its easy to use. They can learn more and better about signal processing and machine diagnostics, due to the implementation of experimental tests that include signal acquisition as they can do the whole process.

The laboratory practice has improved the marks of the students, both in the work group and in the final exam. Besides, they transmitted better satisfaction level in the surveys.

References

1. Kuzlyakina, V.V.: Modern Means and Technologies of Training on Course Theory of Mechanisms and Machines, New Trends in Educational Activity in the Field of Mechanism and Machine Theory, Mechanisms and Machine Science, vol. 19. Springer, Berlin (2015)
2. Chicharro, J.M., Morales, A.L., Nieto, A.J., Pintado, P.: An Example of Inquiry-Based Learning for Undergraduate Mechanical Vibrations, New Trends in Educational Activity in the Field of Mechanism and Machine Theory, Mechanisms and Machine Science, vol. 19. Springer, Berlin (2015)

3. Case Western Reserve University Bearing Data Center Website. https://csegroups.case.edu/bearingdatacenter/pages/welcome-case-western-reserve-university-bearing-data-center-website. Accessed 27 July 2017
4. ECTS User's Guide, European Commission: Supporting and Improving Education and Training in Europe. http://ec.europa.eu/education/ects/users-guide/index_en.htm. Accessed 26 July 2017
5. User Operating Manual for Machinery Fault Simulator, Spectra Quest Inc., Richmond (1996–2003)
6. Gómez, M.J., Castejón, C., Barber, R., Rubio, H., García-Prada, J.C.: BTOOL: a friendly teaching tool to acquire and process vibration signals. In: 14th World congress in Mechanism and Machine Science, Taipei, Taiwan (2015)
7. Zamorano, M.: Guion de prácticas "Ensayos en Máquina de Simulación de Fallos" (Practice guide "Tests in Machinery Fault Simulator), Universidad Carlos III de Madrid (2017)
8. García Prada, J.C., Pedrero Moya, J.I.: Vibraciones en Máquinas. In: Cálculo, Construcción y Ensayo de Máquinas I, 1st edn, Universidad Nacional de Educación a Distancia, Madrid, pp. 44–52 (1993)
9. Feldman, M.: Hilbert Transform Applications in Mechanical Vibration. Wiley, New York (2011)
10. Gómez, M.J., Castejón, C., Corral, E., García-Prada, J.C.: Analysis of the influence of crack location for diagnosis in rotating shafts based on 3x energy. Mech. Mach. Theory **103**, 167–173 (2016)
11. Gómez, M.J., Castejón, C., García-Prada, J.C.: Crack detection in rotating shafts based on the 3x energy. Analytical and experimental analysis. Mech. Mach. Theory **96**, 94–106 (2016)

Dynamics and Mechanical Vibrations. Complementing the Theory with Virtual Simulation and Experimental Analysis

M. Urízar(✉), V. Petuya, A. Hernández, and E. Macho

Faculty of Engineering in Bilbao, Department of Mechanical Engineering, University of the Basque Country (UPV/EHU), Leioa, Spain
monica.urizar@ehu.es

Abstract. The purpose of the present work is to show how the complex theory in subjects related to Mechanism and Machine Theory can be reinforced with some practicals in which the students can perform virtual simulation of the mechanisms under study, and they can even interact with real prototypes to validate by means of experimental analysis the theoretical results. This works deals with Dynamics and Mechanical Vibrations, and presents the capacities of the Dynamics module implemented in GIM software related to the obtaining of free solid diagrams, internal forces maps and diagrams, motion simulation, and so on. In addition to this, interacting with prototypes to carry out experimental measures is also proposed, so that the students can acquire a deeper understanding of some phenomena related to mechanical vibrations.

Keywords: Education · Dynamics · Mechanical vibrations · Virtual simulation
Experimental analysis

1 Introduction

Basing on our own experience [1], specific computer programs constitute a very helpful tool to help students to easily achieve a better comprehension of the theoretical concepts explained in the lectures [2, 3]. Indeed, currently, most of the textbooks related to Mechanism and Machine Theory include simulation programs to reinforce and complement the contents of the book [4, 5].

The subject this work deals with, *Mechanism Theory and Mechanical Vibrations,* plays an essential role in education on Mechanical Engineering. In the Faculty of Engineering in Bilbao, University of the Basque Country (UPV/EHU), this subject is taught during the third course in Mechanical Engineering [6, 7], and it covers significant topics, such as, planar motion geometry, analytic methods, dimensional synthesis, dynamics (inverse and direct dynamics), flywheels, mechanical vibrations, and so on. In general, students encounter some difficulties when facing up to the complex concepts intrinsic to the theoretical bases.

We presented in [1] a work in which virtual simulation in GIM software [3] enabled the students to go deeper into the kinematics of planar mechanisms. Following with this effort, we propose now two main actions to facilitate the understanding of the second

J. C. García-Prada and C. Castejón (Eds.): *New Trends in Educational Activity in the Field of Mechanism and Machine Theory*, MMS 64, pp. 64–71, 2019.
https://doi.org/10.1007/978-3-030-00108-7_8

part of the subject. On the one hand, use GIM software to perform virtual simulation of the Dynamics problem, and, on the other hand, complement the theory of mechanical vibrations with experimental analysis with real prototypes. GIM is a registered software created by the COMPMECH Research Group belonging to the Department of Mechanical Engineering of UPV/EHU. The software is intended for educational purposes. Until now the modules implemented in the software dealt with kinematic analysis, motion simulation and synthesis. Currently, another module, called *Dynamics*, has been incorporated [8] and it is precisely this module the one we propose to reinforce the part of the subject in which the dynamic problem is studied.

Regarding the mechanical vibrations, students carry out some practicals in which they can interact with real prototypes and they can visualize the vibration modes of certain mechanical systems, they can assess the resonance phenomena, they can understand the damping effect, and so on. The main target is to acquire a complete vision of the subject, from theory to practice, which, in our opinion, is the correct way to really understand the subject.

2 Dynamics. Virtual Simulation in GIM Software

Bearing in mind the objective of helping the students to better understand the concepts illustrated in the subject, the theoretical lectures are complemented and reinforced with practical exercises. On the one hand, in the laboratory, virtual simulation of the dynamics behaviour of certain mechanisms is carried out in GIM software. On the other hand, the students, at the time of studying the subject on their own, can freely download the software from the web site of COMPMECH Research Group: http://www.ehu.eus/compmech/software/. In this way, they can check and validate the results they theoretically obtain when solving the dynamics problem.

2.1 Capacities of Dynamics Module

A screenshot of the main window of *Dynamics module* in GIM is shown in Fig. 1. All the modules implemented in GIM offer an easy-to-use format in which the user can navigate through all the existing options and simply select the ones he/she is interested in.

As an illustrative example, a landing gear mechanism of an aircraft has been modelized (see Fig. 1) and the trajectories of points of interest can be visualized. For this first step, the geometry of the mechanism has been already created and its motion has been simulated. Both actions are achieved by using the modules called *Geometry* and *Kinematics*. Students are already familiarized with these modules as they are used in the first part of the subject which deals with kinematics [1].

One of the analysis types that the *Dynamics module* includes is the Inverse Dynamics. It enables the user to obtain the necessary forces or moments to actuate the mechanism once the motion is known and the set of applied loads have been predefined. These applied loads can be punctual forces or moments, as well as linearly distributed forces. As the motion has been already obtained, the acceleration of the gravity centre and the angular acceleration of each moving element is known in all the poses of the mechanism. Then, once the mass properties are defined, the inertial loads are computed.

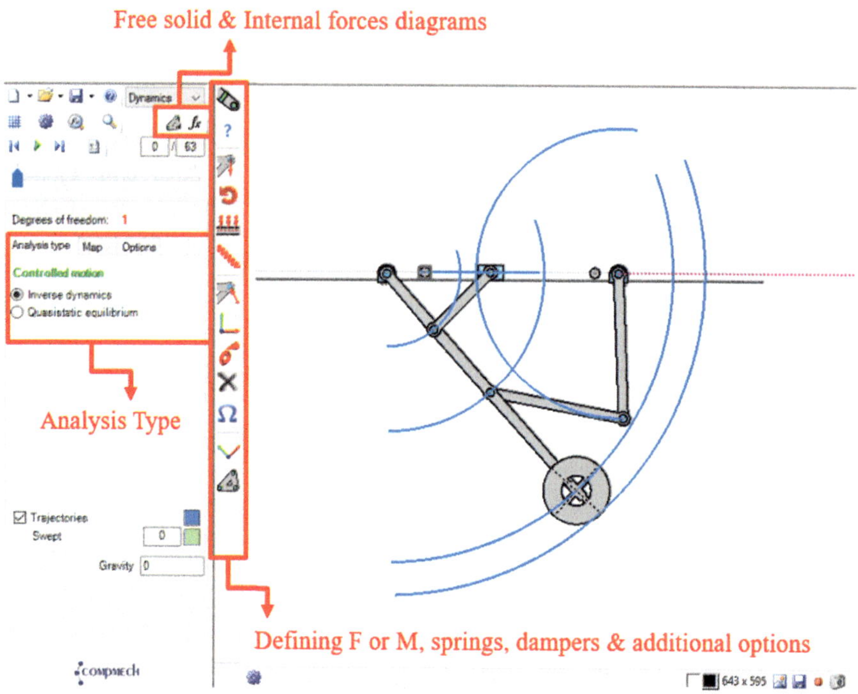

Fig. 1. Main window of Dynamics module in GIM software

Basing on D'Alembert principle, in which the inertial loads are treated as external virtual loads, the free solid diagrams of any element of the mechanism can be obtained. As an example, Fig. 2 shows the free solid diagram of one of the bars of the landing gear mechanism. It is important to emphasize that these diagrams can be obtained for any pose of the mechanism. In this way, the students can visualize the evolution of the diagrams along the motions of the mechanism. Another analysis type implemented in the module is the one called Quasi-static equilibrium. It works in a very similar way to the Inverse Dynamics with the difference that in the quasi-static one the inertial loads are considered null.

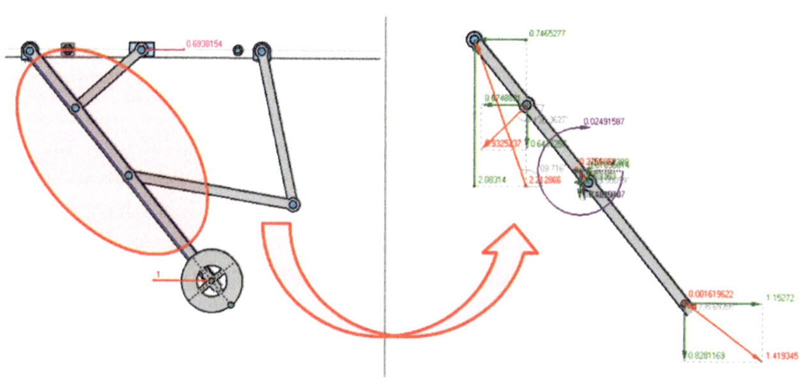

Fig. 2. Free solid diagrams

Additionally, the internal forces of any element of the mechanism can be computed. Two main options are offered to the user. The first one is to visualize the maps of the internal forces of all the elements of the mechanism. This is shown in Fig. 3a, in which coloured maps associated to the axial forces, the shear forces and the bending moments are plotted. The second option is to represent the diagrams of the internal forces for a selected element, as it is depicted in Fig. 3b.

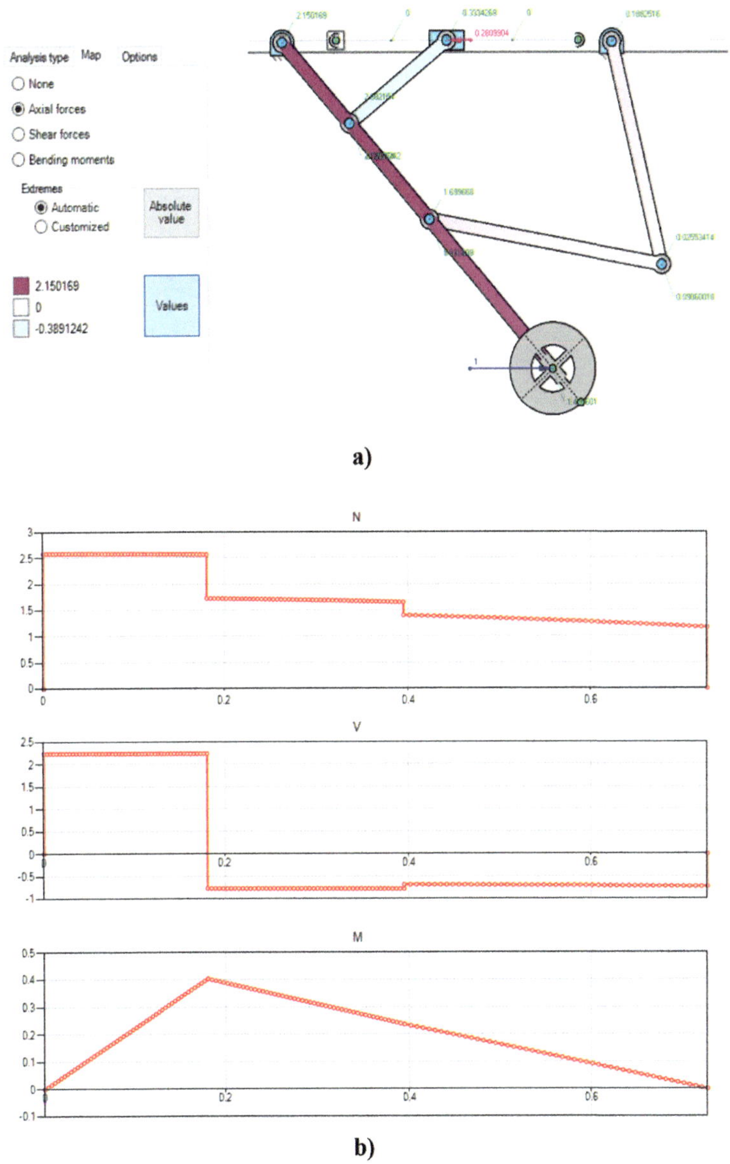

Fig. 3. Internal forces maps and diagrams

Dynamics module also implements another analysis type called Direct dynamics. The target of this approach is to obtain the motion of the mechanism once the set of actuating and applied loads is predefined and some initial conditions are established. In this case, the user only defines the geometry of the mechanism under study (in *Geometry module*) and then, once inside *Dynamics module*, selects the option Direct dynamics. The user then defines the set of loads applied to the mechanism, the initial conditions of position and velocity, and the software automatically computes the corresponding motion. An illustrative example is shown in Fig. 4, where the motion of a bar connected by springs to the fixed element and subjected to its own weight can be visualized.

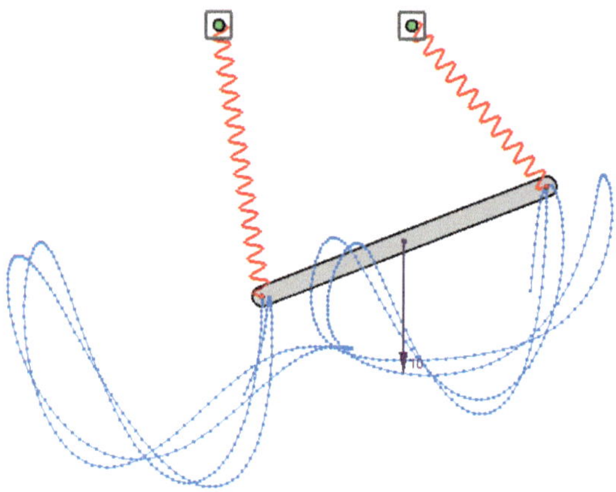

Fig. 4. Motion simulation in Direct dynamics

3 Vibration Experimental Analysis Interacting with Prototypes

The second part of the subject *Mechanism Theory and Mechanical Vibrations* deals with the phenomena of free or forced mechanical vibrations of single and multiple degree-of-freedom mechanical systems. In relation to forced mechanical vibrations, the response of the system to harmonic type excitations, as well as the response to step, pulse and transient vibration excitations and their combinations is studied. Modal analysis technique is also studied for multiple dof mechanical systems.

Apart from this lengthy theory, some concepts regarding the experimental analysis of mechanical vibrations are explained during the lectures. Additionally, the theoretical concepts are reinforced with some practicals in the lab, in which the students interact with real prototypes and carry out experimental measures. To cite some examples, the students visualize the vibration modes of a beam making use of a stroboscope, they obtain the natural frequencies of a prototype of a building using an impact hammer and analyzing the response measured by the accelerometers, they assess the vibration modes of the prototype being subjected to a harmonic force varying its frequency and they analyse the resonance phenomenon of a single dof mechanical system.

One of these aforementioned practices is shown in Fig. 5. In this Fig. 5a, the prototype of a single dof mechanical system actuated by a slider-crank mechanism driven by an electric motor is displayed. In this case the vibrations are induced to the mass due to the motion imposed on the slider. The students obtain experimentally the displacement of the mass as a function of the frequency of the input (the angular velocity of the crank) and confirm the high amplitude the mass acquires when getting close to the natural frequency (see Fig. 6). These results are validated with the ones they obtain by developing the theoretical equations, and they are also validated with the simulation of the mechanical system using GIM (Fig. 5b).

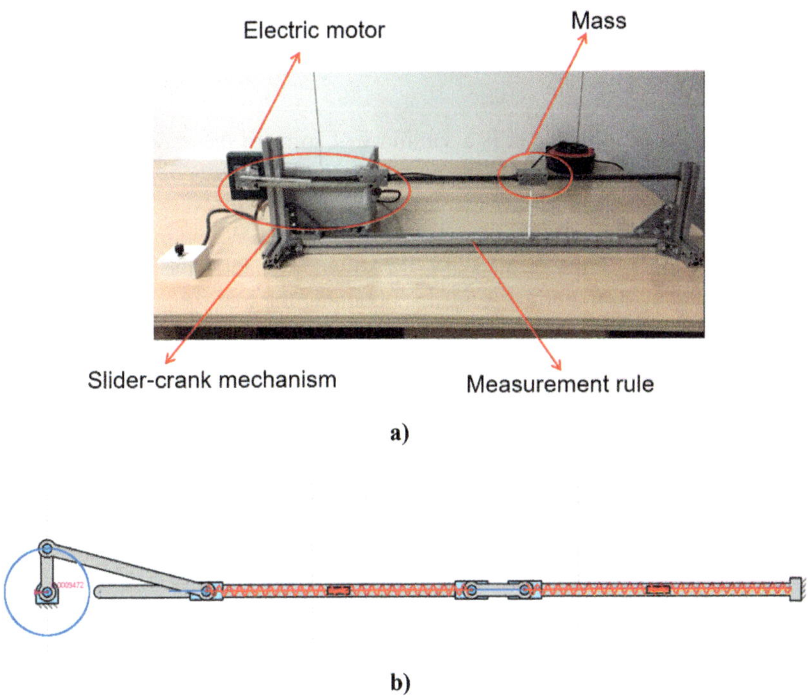

Fig. 5. Mechanical vibrations: (a) prototype and (b) simulation

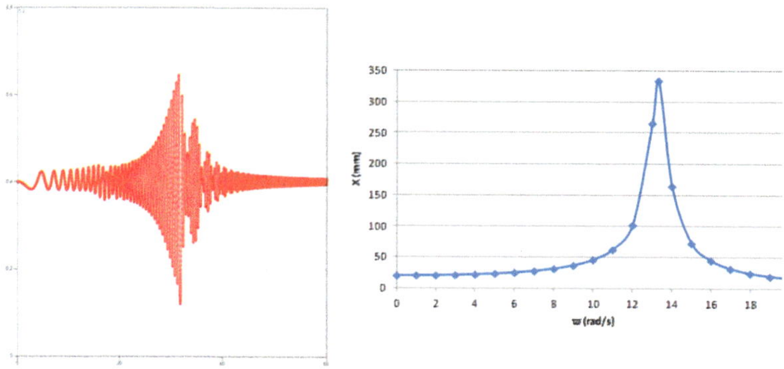

Fig. 6. Validating experimental results

4 Conclusions

The work presented in this paper is a summary of some of the actions proposed by the teaching staff of the subject *Mechanism Theory and Mechanical Vibrations* with the purpose of facilitating the students a deeper comprehension of relevant concepts, in particular, the ones related to Dynamics and Mechanical Vibrations. On the one hand, the Dynamics module implemented in GIM software enables the students to perform simulations of many different mechanisms, obtaining the free solid diagrams and maps, the values of the actuating loads, the motion of the mechanism under certain loads. On the other hand, mechanical vibrations theory is complemented with experimental analysis so as to understand the insights of important concepts, such as resonance, damping effects, vibration modes, and so on. Moreover, the combination of simulation in GIM and experimental analysis becomes a strong tool for the students when facing up to the subject.

Acknowledgments. This work was supported by the Spanish Government through the *Ministerio de Economía y Competitividad* (Project DPI2015-67626-P (MINECO/FEDER, UE)), the financial support from the University of the Basque Country (UPV/EHU) under the program UFI 11/29 and the support to the research group, through the project with ref. IT949-16, given by the *Departamento de Educación, Política Lingüística y Cultura* of the Regional Government of the Basque Country.

References

1. Urízar, M., Altuzarra, O., Diez, M., Campa, F., Macho, E.: Kinematic analysis of planar mechanisms by means of examples. In: García Prada, J.C., Castejón, C. (eds.) New Trends in Educational Activity in the Field of Mechanism and Machine Theory. Springer, Berlin (2014)
2. Pennestrì, E., Vita, L.: Multibody dynamics in advanced education. In: Advances in Computational Multibody Systems. Springer, Dordrecht, pp. 345–370 (2005)

3. Petuya, V., Macho, E., Altuzarra, O., Pinto, Ch., Hernández, A.: Educational software tools for the kinematic analysis of mechanisms. Comput. Appl. Eng. Educ. (2011). https://doi.org/10.1002/cae.20532
4. Waldron, K.J., Kinzel, G.L.: Kinematics, Dynamics and Design of Machinery. Wiley, New York (1999)
5. Norton, R.L.: Design of Machinery: An Introduction to the Synthesis and Analysis of Mechanisms and Machines. McGraw-Hill, Boston (2003)
6. Hernández, A.: Cinemática de mecanismos: Análisis y Diseño. Editorial Síntesis, Madrid (2010)
7. Hernández, A., Pinto, Ch., Agirrebeitia, J., Petuya, V.: Dinámica de Máquinas. Publicaciones dela Escuela de Ingeniería de Bilbao, Bilbao (2013)
8. Macho, E., Petuya, V., Urízar, M., Hernández, A.: Software de propósito general para resolver la dinámica inversa y calcular los esfuerzos internos de mecanismos planos, 13º Congreso Iberoamericano de Ingeniería Mecánica, CIBEM2017, Lisboa, Portugal (2017)

Innovative and Multidisciplinary Teaching Through the Design and Construction of Low Consumption Vehicles for International Competitions

Massimiliana Carello(✉)

Politecnico di Torino - Mechanical and Aerospace Engineering Department, Corso Duca degli Abruzzi N.24, 10129 Torino, Italy
massimiliana.carello@polito.it

Abstract. The paper presents an innovative way to teach point out at the Politecnico of Torino thanks to the creation, ten year ago, of the Team H_2politO. The interdisciplinary Team, made by students, give an opportunity to tackle a "new teaching", different from the traditional one, which provides a lecturer in classroom, more innovative with protagonist students able to cope with technical problems too complex, such as those who, from the moment of graduation, will have to face the business world. The challenge of the Team is design, build and assembly a working vehicle prototype to participate to an International student competition (Shell Eco-marathon) with a goal: global efficiency to obtain low consumption results.

1 The Team Mission

The Team H_2politO born in 2007 with thirteen students (Fig. 1) and then year after is composed by, more or less, fifty students (Fig. 2).

The Team H_2politO is interdisciplinary and made of students, from bachelor and master degree, which come from different engineering field: mechanical, automotive, aerospace material science, electronic, mechatronic, management.

A Company Team and the students Team (Fig. 3) organization could be compared, where instead of Product and Market there are the Vehicle prototype and the Competition.

The students, thanks to the knowledge and skills acquired during their academic career (Didactic path in Fig. 4), have the opportunity to learn more competencies, also multidisciplinary, managing a real project from A to Z, working together and not individually, solving real problems, typical of the job life (Innovative path in Fig. 4).

The final goal of the Team is take part, with their vehicles, at a competition, in this case the European Shell Eco- marathon, where more than 200 Team (from the best University and Secondary School) are challenging with their vehicles in

J. C. García-Prada and C. Castejón (Eds.): *New Trends in Educational Activity in the Field of Mechanism and Machine Theory*, MMS 64, pp. 72–79, 2019.
https://doi.org/10.1007/978-3-030-00108-7_9

Fig. 1. The first Team in 2007

Fig. 2. The last Team in 2017

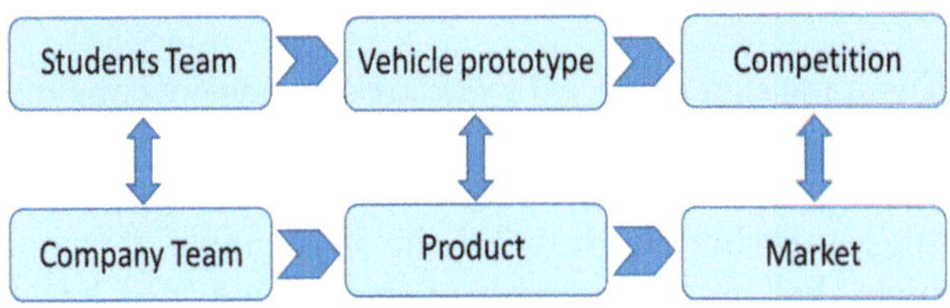

Fig. 3. The students and company comparison

a race of low fuel consumption, in fact the winner is the vehicle that obtain a less consumption.

Observing Fig. 4 the blue narrows explain the work flow while the yellow narrows explain the closed loop of the path, in particular take into account the com-petition results obtained of the Team it is possible to change, year by year, the innovative and the didactic path.

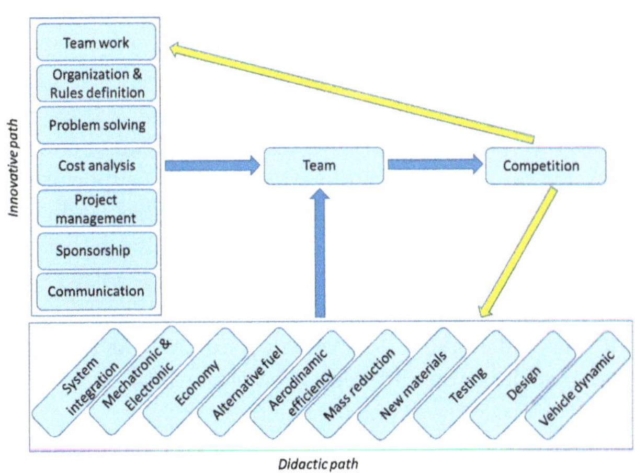

Fig. 4. Team path

The Team H_2politO is a different, innovative and somehow unique project, is not just a Team but also something more: it is a new type of conceiving educational, professional and personal growth. Team members aim at being perceived as an experimental laboratory where competences, capabilities and potentialities of futures engineers are fostered. Students strive to become not only solid and advanced technical experts but also, equally important, down-to-earth managers having excellent communication, leadership and teamwork skills.

Practical and hands-on experiences are doubtlessly a complementary and enriching form of educational path. Therefore, H_2politO firmly believes that learning by doing represents an absolute advantage Team members can count on.

2 The Challenge

The results of Team passion and hard work are low-energy consumption vehicles, in fact, the final goal is to take part and win at the Shell Eco-marathon SEM a low energy consumption cars competition. The SEM is, first of all, an educative project that brings together the value of the sustainable development with the driving of a vehicle that uses an amount of energy that is as low as possible. An international event for students of University and secondary schools, that every year involves more than 200 Teams, and 3000 students in the European version, arriving from all over Europe. The SEM is an international context that helps the comparison between students of all the more important Universities.

To participate at the SEM it is mandatory, first of all, to design the vehicle take into account the rules of the competition, that are relative to dimensions, mass, mechanical and electronic security aspects, etc. and to pass all the selection phases.

The SEM has two different vehicle categories: Prototype (three wheel vehicle) and Urban Concept (four wheel vehicle), and different propulsion system: Battery-electric, Hydrogen and Combustion engine (gasoline, diesel, ethanol,..)

3 The Vehicles

The vehicles of Team H_2politO are always completely design, construct, assembly and test on track by the students under the technical supervision of a Faculty Ad-visor. From 2007 to now 5 different vehicle have been made: 4 hydrogen fuel cell prototypes (From Figs. 5, 6, 7 and 8) and 1 parallel hybrid urban concept (Fig. 9). From 2008 to 2010 the Team had participate with a prototype while from 2011 to now had participate with a prototype and an urban concept [1–16]. When the Team does not make a new vehicle year by year the students improve the characteristics working on different area, take into account problems and goals.

The steps of the activity for the Team are:

1. Analysis of the rules and constrains of the competition (maximum dimensions and mas, security aspect from mechanical and electronic point of view).

Fig. 5. IDRA08 - 2008

Fig. 6. IDRA09 - 2009

Fig. 7. IDRApegasus - 2012

Fig. 8. IDRAkronos 2016

Fig. 9. XAM - 2011

2. Parallel activities of: CAS of the external shape of vehicle and aerodynamic simulation; CAD, force and torque calculation that are exerted in all the components (also making some design hypothesis); FEM model and simulation of the body and all mechanical components; vehicle dynamic and power-train modeling; internal layout study. All the design take into account new technical and innovative solutions and new materials to reduce mass and guarantee the safety of car and driver. Some iterations are mandatory among the different activities to obtain the best result.
3. "Freezing" of the vehicle, final control of all the virtual models results and construction of the components, cooperating with company when it is not possible to realize the components inside the University.

4. Assembling of the vehicle and testing in laboratory if possible (electronic control board, power-train, mechanical transmission, etc.).
5. First "functionality tests on track to have a reliable vehicle and then consumption tests to validate the models.
6. Participation at the competition and application of the race strategy model to obtain the best results.

If the vehicle is not a new one the steps similar, without, for example, the design activities about the body.

4 The Results at the Competitions

Year by year the Team H_2politO is grow up, not only in term of number of students but also in terms of the competition results.

The last results obtained during the 2017 (Fig. 10) are:

- IDRAkronos at Shell Eco-marathon London - 2° place hydrogen prototype category - 831 km/m^3 (of hydrogen in normal conditions) that corresponds to 2467 km/L (of equivalent gasoline).
- IDRAkronos at Shell Eco-marathon Le Mans - 1° place hydrogen prototype category - 1022.21 km/m^3 (of hydrogen in normal conditions) that corresponds to 3034.86 km/L (of equivalent gasoline).
- XAM at Shell Eco-marathon London - 4° place ethanol urban concept category - 10° place for general urban concept combustion engine category - 110 km/L (of equivalent gasoline).

Fig. 10. The Team H_2politO winner at the SEM 2017

But an additional challenge of the Team is to win the off track awards during the SEM, and during this ten year has been obtained: 5 Communication Award: 2009, 2010, 2012, 2014, 2017 and 3 Design Award: 2010 (with IDRA), 2011 (with XAM) and 2016 (with IDRAkronos).

For the Team to win a Design Award does it means to have made a high level of global design of the vehicles while to win the Communication Award does it mean to have made a very good work "all around" to involve other students and companies.

5 The Results for the Students

The Politecnico of Torino is the only University in Italy that recognize the Team activity in terms of credits. In fact, after a period of work inside the Team for the students, in agreement with the Faculty Advisor, is possible to obtain:

- 6 ECTS credit with "Team learning activities in student" instead of a choice exam during the bachelor;
- 6 or 12 ECTS credit with "Team learning activities in student" instead of a choice exam during the master.

However, if the student stays inside the Team more than one year has the possibility to "transform" their work in a thesis. Until now inside the Team more than 150 thesis have been developed, and this is a very innovative educational aspect of the Team. In fact, they allow the realization of the thesis cluster that combines technological, organizational areas of the development H_2politO has envisioned and embraced, and that allow the important transmission, year by year, the knowledge at the students, from the past to the future.

But for the students it is possible to write technical papers for journals and/or participate at International Conference, showing the high technical level of their vehicles in the different areas [1–16].

6 Conclusions

The Team believes in hard work as the basis of future success. Students crave for continuously improving and strive for exceeding expectations by nurturing the team spirit in order to create those synergies able to add value to individual performances and capabilities. Therefore, passion and team spirit are really the foundation of H_2politO values.

The work at the Team and the prototype can be perceived as an business laboratory where is required a synergy between technical knowledge and managerial skills. Therefore, H_2politO allowed developing both technical competencies strictly linked to the Prototype/Product and managerial competences characteristics of a business team.

The Team believes that the competition is a way to achieve excellence. The market, thus the competition, gives them the possibility to work hard in order

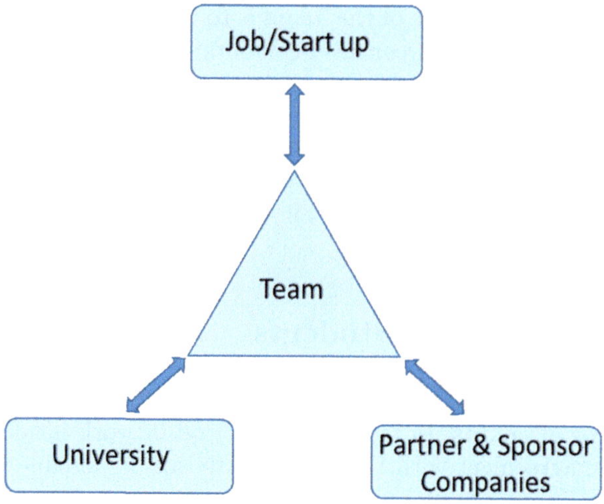

Fig. 11. The innovative rule of the Team H_2politO

to challenge the competitors and produce a more advanced technology vehicle that could become benchmark for its category.

The Team (Fig. 11) wants, through the University to drive innovation into everything he does. It works to become a mean by which partners and sponsors could create and develop innovative path to prepare the future engineers at the job, in a company or to create new innovative start up.

References

1. Airale, A., Carello, M., Scattina, A.: Carbon fiber monocoque for a hydrogen prototype for low consumption challenge. Materialwissenschaft und Werkstofftechnik **42**(5), 386–392 (2011)
2. Brusaglino, G., Buja, G., Carello, M., Carlucci, A.P., Onder, C.H., Razzetti, M.: New technologies demonstrated at formula electric and hybrid Italy 2008. World Electr. Veh. J. **3**(1), 160–171 (2009)
3. Carello, M., Airale, A., Ferraris, A., Messana, A.: XAM 2.0. from student competition to professional challenge. Comput. Aided Des. Appl. **11**, 61–67 (2014)
4. Carello, M., Airale, A.G.: Composite suspension arm optimization for the city vehicle XAM 2.0. Adv. Struct. Mater. **54**, 257–272 (2014)
5. Carello, M., Airale, A.G., Ferraris, A. City vehicle XAM 2.0: design and optimization of the composite suspension system. SAE Technical Papers, 1 (2014)
6. Carello, M., Airale, A.G., Messana, A.: IDRApegasus: a carbon fiber monocoque vehicle prototype. Materialwissenschaft und Werkstofftechnik **45**(5), 397–405 (2014)
7. Carello, M., Andrea, S., Airale, A.G., Ferraris, A.: Design the city vehicle XAM using CFD analysis. SAE Technical Papers, April 2015
8. Carello, M., Bonansea, P., D'Auria, M.: Driveline optimization for a hybrid electric city vehicle to minimize fuel consumption. SAE Technical Papers, 1 (2014)

9. Carello, M., De Vita, A., Ferraris, A.: Method for increasing the humidity in polymer electrolyte membrane fuel cell. Fuel Cells **16**(2), 157–164 (2016)
10. Carello, M., Ferraris, A., Airale, A., Fuentes, E.: City vehicle XAM 2.0: design and optimization of its plug-in E-REV powertrain. SAE Technical Papers 1 (2014)
11. Carello, M., Filippo, N., D'Ippolito, R.: Performance optimization for the XAM Hybrid Electric Vehicle prototype. SAE Technical Papers (2012)
12. Carello, M., Messana, A.: IDRApegasus: a fuel-cell prototype for 3000 km/L. Comput. Aided Des. Appl. **12**, 56–66 (2015)
13. Carello, M., Scattina, A.: Structural design and experimental investigation of a carbon fibre wheel for low consumption vehicle. Adv. Struct. Mater. **16**, 521–535 (2012)
14. Cubito, C., Rolando, L., Ferraris, A., Carello, M., Millo, E.: Design of the control strategy for a range extended hybrid vehicle by means of dynamic programming optimization. pp. 1234–1241 (2017)
15. Filippo, N., Carello, M., D'Auria, M., Marcello, A.: Optimization of IDRA pegasus: fuel cell hydrogen vehicle. SAE Technical Papers, 2 (2013)
16. Xu, S., Ferraris, A., Airale, A.G., Carello, M.: Elasto-kinematics design of an innovative composite material suspension system. Mech. Sci. **8**(1), 11–22 (2017)

Experimental Determination of the Fundamental Parameters of Road Traffic and Noise Levels Emitted

S. Sanz(✉), V. Díaz, M. J. Boada, and B. Boada

Mechanical Engineering Department, Universidad Carlos III de Madrid, Avda. de la Universidad, nº30, 28911 Leganés, Spain
ssanz@ing.uc3m.es

Abstract. In this work the main objective is transmit experimentally to the student the basic knowledge about the parameters of the traffic engineering in an easy, simple and effective way. Nowadays are used different computer softwares that allow to calculate the basic parameters who characterize the traffic. However, the teaching experience over the years, has allowed us to know that a student is capable of fixing the knowledge in a better way when he learns by practical procedures. That's why a simple practice has been designed, where the students collect data about the appraisals of the rolled traffic that circulates along a route section near to Leganes's Campus (University Carlos III of Madrid), and then they analyse and calculate the typical parameters of the traffic, fixing this way the engineering of the traffic concepts (There has been chosen an urban section route that meets the necessary requirements to determine the basic parameters of traffic.).

The accomplishment of this practice is going to allow to know the problematics that supposes a real study of the principal characteristics of the traffic: intensity, velocity, density… parameters that are used to obtain a better knowledge of the behaviour of the traffic. In reality these studies require certain investments in equipments who suppose an economic cost. In the practice, since there is a limited team of pupils, only a chronometer and a metric tape is needed. It is enough to realize a number of approximations to conclude the value of this traffic parameters.

Also, by using sonometers the urban noise is registered, allowing to calculate a number of indexes that will allow to evaluate the sonorous level existing in the zone.

1 Introduction

The Traffic Engineering is the branch of the Engineering that has like principal objective the planning, tracing and exploitation of the road networks, so that the persons and goods traffic is sure, rapid and economic.

In the objectives identified they can separate two aspects; the planning road networks and the arrangement and regulation of the traffic in existing networks.

For this case, is studied the continuous traffic which is defined as the type of traffic in which external elements of regulation of the traffic do not exist, such as traffic lights,

J. C. Gacría-Prada and C. Castejón (Eds.): *New Trends in Educational Activity in the Field of Mechanism and Machine Theory*, MMS 64, pp. 80–88, 2019.
https://doi.org/10.1007/978-3-030-00108-7_10

which force to stop the vehicles. The conditions of traffic are the result of the interactions between the vehicles, and between these and the geometric and environmental characteristics of the road.

Given the complex of the traffic, exist multitude of variables that influence in his behaviour. Three of them are considered the fundamental ones: intensity, density and speed.

1.1 Intensity of the Traffic

It is named intensity of traffic to the number of vehicles that pass through a certain cross section of a road for unit of time (Fig. 1).

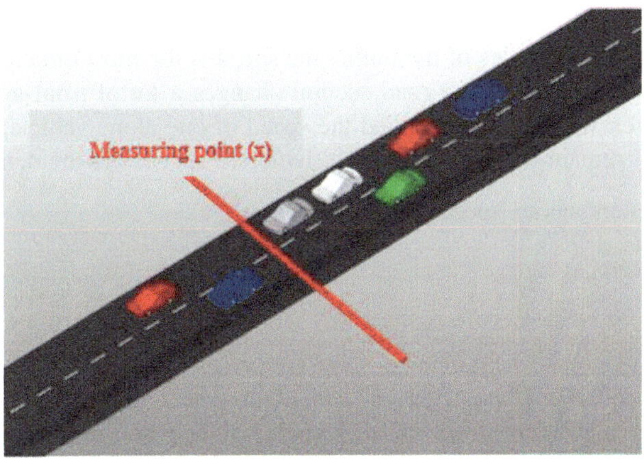

Fig. 1. Scheme of determination of the intensity

$$I = \frac{n(x)}{t}$$

where:

$n(x)$: number of vehicles that pass through the fixed cross section (x).
t: temporary studied interval (seconds, hours, days, etc.).

Depending on this temporary interval, the intensity has one or another unit. Most used are vehicles/hour and vehicles/day. When one uses as unit vehicles/hour speaks about hourly intensity, and when it is used vehicles/day speaks about daily intensity.

Generally, the concept of intensity is applied to temporary intervals equal or superior to an hour. For this practice the time is 30 min, so it must use the equivalent variables for the traffic volume and hourly intensity.

- Traffic volume: it is the number of vehicles that cross a certain section of road during a certain time.

- Hourly equivalent intensity: it is defined as the quotient between the number of vehicles that cross a determined section during a period of time lower than an hour and the temporary used interval expressed in hours.

With these variables it is possible to approximate the intensity of the traffic in an hour with appraisals of a minor time. The obtained value is not any more than an estimation, because the intensity changes in the time and therefore it does not have to be uniform throughout the considered hour.

The hourly intensity is used for the aspects linked to the project and the management of the roads: capacity of the roads, characteristics of the intersections and links, traffic control, coordination of traffic lights and management of the traffic.

1.2 Velocity

Of the fundamental variables of the traffic, the speed is the most difficult to determine. It is evident that the speed in a road section changes a lot of from a one vehicle to another, but even when there is studied the speed of one alone vehicle, it can be seen that it does not remain constant over time. Different speeds can be defined:

- Local or instantaneous speed
- Running speed
- Tour speed
- Section average speed
- Average speed in movement

Provided that the practice develops in an urban zone and there is not equipment to know the speed of each one of the vehicles, there will be requested the average speed of the studied section.

Being the quotient between the length of a section and the average time used by a group of n vehicles in crossing it.

$$V_r = \frac{L}{\frac{\sum_{j=1}^{n} t_j}{n}} = \frac{L \cdot n}{\sum_{j=1}^{n} t_j}$$

where:

n: Number of observed vehicles
t_j: Time that the vehicle j has used in crossing the section
L: Length of the section.

1.3 Density

There is understood by density of traffic the number of vehicles that take a place in a road section of length given in a certain instant (Fig. 2). It is usually measured in vehicles/km. The maximum value of the density takes place when all the vehicles are in row without gaps between them and logically it depends on the average length of the vehicles (and, therefore, of the composition of the traffic in every moment).

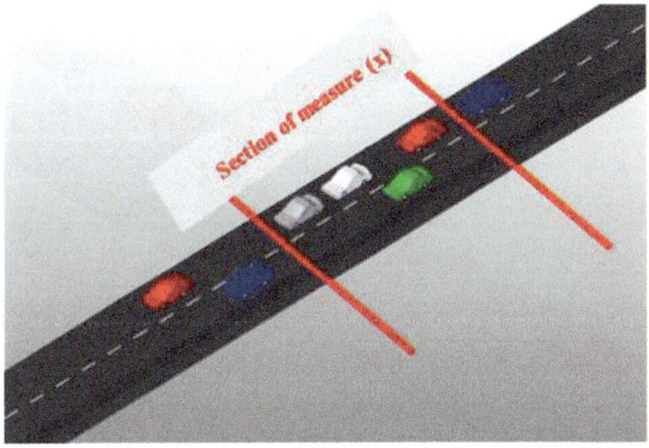

Fig. 2. Scheme of determination of the density

Nevertheless, it is possible to calculate across the section average speed and the intensity of traffic, that they are simpler to measure. Being the three fundamental variables linked by the fundamental formula of the traffic:

$$I = D \cdot V$$

There are other parameters that linked to the previous ones that will be object of the practice, as for example, the calculation of the average spacing, the interval or the capacity of the road.

1.4 Spacing

The spacing between vehicles is defined as the distance between the frontal parts of a vehicle and the one that follows in the same lane in an instant (Fig. 3).

The average value of this variable is the inverse of the density.

$$s_m = \frac{1}{D}$$

For a certain speed, there can be defined also a separation or average minimal spacing that guarantees the safety of the traffic.

For the calculation of the average separation, s_m, it is considered the real conditions of traffic of the road, and what is obtained is the real average separation of the considered vehicles. One must not confuse the s_s with the s_m. It must be taken into account that, s_s represents the spacing that, for average term, the vehicles should guard to guarantee that the traffic develops in safety conditions. This average separation of safety does not have to match with the average real separation of the vehicles.

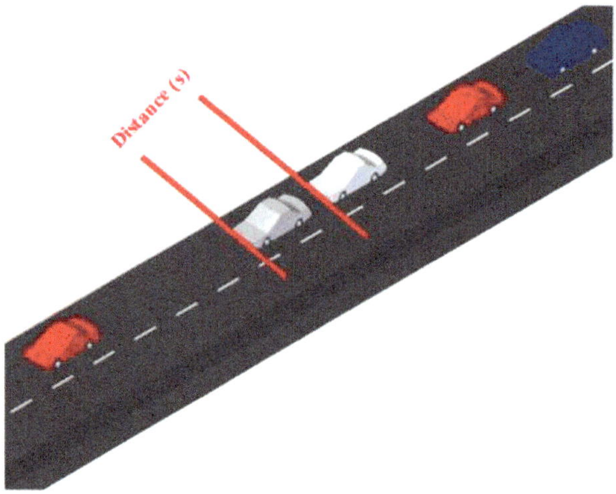

Fig. 3. Spacing between vehicles

To calculate the safety separation between vehicles there are a series of mathematical formulas and, in general, they consist of three addends:

- The First addend: Constant, function of the length of the vehicles.
- Second addend: Linear depending on the speed, function of the space crossed during the time of perception. This term will depend on the reaction time of the driver.
- Third addend: Quadratic depending on the speed, it includes the additional effect of the stopping.

$$s_s = a + bV + cV^2$$

According to the values that are given to these parameters a, b and c, different relationships will be obtained.

1.5 Interval

The time passed between the pass of two successive vehicles for the same section is named an interval, τ.

The average value of the intervals measured for diverse vehicles during an interval of time is intimately linked to the intensity according to the following equation:

$$\tau_m = \frac{1}{I}$$

Using the fundamental equation of the traffic we link the variables intensity, velocity and density. Supposing that two vehicles move on an average speed V_t separated a distance s, the interval between both vehicles is not any more than:

$$\tau = \frac{s}{V_t}$$

From this relationship and from the definitions of interval and spacing, it is had:

$$\left.\begin{array}{l} \tau = \frac{s}{V_t} \\ s = \frac{1}{D} \\ \tau = \frac{1}{I} \end{array}\right\} \Rightarrow \frac{1}{I} = \frac{1}{D \cdot V_t} \Rightarrow I = D \cdot V_t$$

If the traffic was homogeneous, that means, all the vehicles were circulating at the same speed and keeping the same separation with the previous vehicle, the previous reasoning would be enough to establish the relation between the three fundamental variables.

1.6 Capacity

The capacity of a lane, expressed in vehicles/hour, is the maximum number of vehicles that it can cross through it for unit of time. In agreement with this definition, the capacity of a road always has to be equal or bigger than his intensity.

Supposing a uniform speed, the capacity can be expressed like:

$$C = 1000\frac{V}{s_s}$$

where V is the speed in km/h and s_s is the minimal average separation between vehicles calculated according to one of the expressions seen in previous sections.

1.7 Measurements of Noise

The noise sources in a vehicle are diverse and are distributed along the vehicle. In addition, they are going to be a function of variables as the speed and the condition of the road, among others. For a fixed observer, the cross of a vehicle generates a level of ascending noise as it approaches the point where the observer is, being after his cross a descending trend. The perception is going to be different depending on the type of vehicle, the type of conduction, and the environment where it is circulating. In the study of the traffic noise, it is necessary to take into account not only the vehicle on an individual way, but it will be necessary to combine the theory of the noise to an individual vehicle and the theory of the flow of the traffic. In the study of the noise produced by a vehicle only it is taken to account the time during the sonorous level overcomes the background noise, without thinking what happens before and after the cross of the vehicle. In case of the measurement of the traffic noise, the parameters that influence the traffic complicate the possibility of describe it mathematically, it means, the relations that are established in a condition are only valid for this situation. For it, the best way of describing the noise in the traffic is by probabilistic techniques.

For this case, there is going to be used the Sonorous Equivalent Level, L_{eq}. To the be a level that measures the total energy in a period of time, it represents with just one number this energy, being able to ignore the random character of the traffic. Also, the different sources contributions in the same instant can be added logarithmically to give the equivalent total level.

The equivalent constant level is the level in decibels of a hypothetical constant signal corresponding to the same quantity of acoustic energy than the vibrant real signal considered during the same period of time T:

$$L_{eq} = 10 \log \frac{1}{T} \int 10^{\frac{L(t)}{10}} dt$$

$$L_{eq} = 10 \log \frac{1}{T} \sum_{i} 10^{\frac{L_i}{10}}$$

2 Development of the Practice

For the accomplishment of the practice they are going to take a series of measures in the proximities of Leganés's Campus. These measures will include:

- Appraises
- Speed
- Noise

In order to obtain the largest information of the rolled urban traffic in the surroundings of the Campus, the students will be distributed in several groups of practices, which will be located according to the positions that appear in the map (Fig. 4) and they will take measured simultaneously.

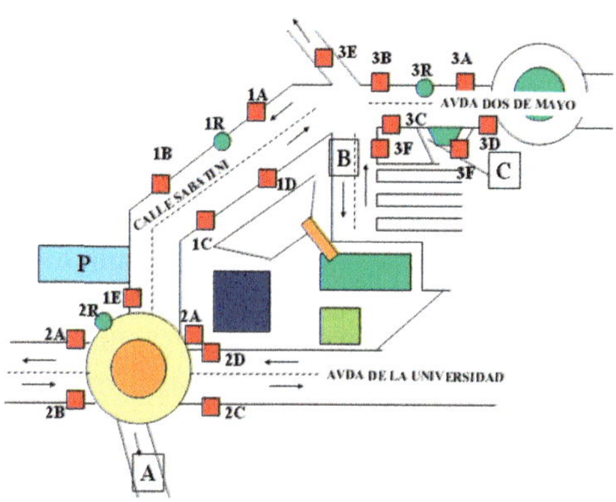

Fig. 4. Map of the zone of the accomplishment of the capture of measures

The capture of information in the designated position will be 30 uninterrupted minutes.

Likewise, it is very important that all the positions that measure during the same session do it during the same 30 min.

As for the measure of appraisals, there will be grouped the vehicles counted in groups of 5 min and distinguishing them according to his nature (tourisms, industrial vehicles and motorcycles).

To know the speed of the vehicles, it is necessary to control the time that the vehicles late in crossing a certain section of route. The length of the above mentioned section will be choosing by the designated group, being like minimum 50 m.

As for the measures of urban noise, there will be registered measures of 1 min distributed throughout the 30 min in which the information of appraisals will be taken.

To measure the different levels of sonorous pressure it has three sonometers. The measurements of noise will be done by the following form:

- Scale of weighting A.
- Continue level equivalent to 1 min.
- Integration speed: FAST.
- Prevent the registered measures from leaving the scale.

As soon as he gets the information of all the positions that realize the capture of information during the same session of practices, it is asked:

(1) Determine for every type of vehicle and for every position of measure, the following directional intensities: I5MAX, I5MEDIA, I15MAX and I30. Comment the results obtained.
(2) Determine for every type of vehicle, in the direction in which they have placed: the average speed of the section.
(3) Calculate for the traffic in his set, without distinguishing by type of vehicle, the average local speed and the average speed of the section. Compare the results with the obtained ones in the previous part and comment the results.
(4) Determine the average value of the intervals.
(5) Determine the density in those sections.
(6) Determine the average separations of *a* and *s*. The coefficients are b = 0.2 and c = 0.003. Take length of motorcycle: 2 m, length of tourism: 4 m and length of heavy vehicle: 8 m.
(7) Being based on the obtained results, estimate the level of service, according to his judgment, of the both analyzed roads.
(8) Determine for each of the positions of measure of noise:
    - The maximum levels of noise registered.
    - Equivalent results of constant level.
(9) Does correlation exist between the parameters of traffic with those of acoustic contamination?
(10) Economic valuation of the study of appraisals realized considering staff, duration, material used, etc.

3 Conclusion and Future Work

After the theoretical explanation in the classroom and the application of the fundamental concepts in the practice, the students will achieve better knowledge by a not conventional way, realizing a fieldwork in the urban road. In addition, at the moment of the experimental capture of measures, you can observe the effort to be correct and to guarantee a study of the traffic coherent.

For answers in the questionnaires of different groups, the application of the theoretical concepts makes the students see the usefulness of the same ones in the professional world.

As a future work it is studying to realize the practice in discontinuous traffic, realizing the capture of measures in a roundabout with three possible entrances and exits.

Reference

1. Díaz, V., et al.: Transportes, pp. 375–540. UNED, Madrid (2006)

MMS in the Engineer Program: Applications and Research

A Comparative Analysis of Teaching MMS at Politehnica University of Timişoara and University of Cassino and South Latium

Erwin-Chr. Lovasz[1(✉)], Marco Ceccarelli[2], Corina Mihaela Gruescu[1], Giuseppe Carbone[2], and Iosif Cărăbaş[1]

[1] Politehnica University of Timişoara, Bv. Mihai Viteazu 1, 300222 Timişoara, Romania
erwin.lovasz@upt.ro

[2] University of Cassino and South Latium, via G. Di Biasio 43, 03043 Cassino, Italy
ceccarelli@unicas.it

Abstract. Teaching activities in MMS field has a long tradition at both Polytehnica University of Timişoara and University of Cassino and South Latium. The paper shows a comparative study regarding the evolution of teaching mechanism theory with the curricula, representative professors and teaching materials in both universities, with challenges for the future.

Keywords: History of mechanism and machine science · Teaching activity Teaching materials · Professors

1 MMS Teaching in IFToMM Frames

Mechanism and Machine Science (MMS) education has been and still is one of the main aspects of IFToMM activity (see www.iftomm.net). In fact, a Permanent Commission on Education is active since the early days in IFToMM, since it was established at the foundation of IFToMM in 1969.

Teaching MMS is worked out worldwide as both fundamental and advanced part of a modern formation of engineers not only in mechanical engineering. The core of mechanical systems is based on the mechanics of rigid bodies, in terms of motion and dynamic actions, and therefore, this teaching is planned everywhere in the world at the beginning of formation programs, even if in different ways and extents. Several textbooks are published in many languages and they are the references of MMS not only in IFToMM community. In addition, advanced topics are programmed in the curriculum of mechanical engineering as referring to specific problems, systems and technologies as part of specialization and multi-disciplinary skills. Mainly in these advanced topics IFToMM provides international frames for collaboration. They are the Technical Committees (TC) working on specific areas or subjects. In terms of teaching, TCs are active not only to share and plan common teaching as part of the mission of directing to new frontiers, but they also organize specific teaching in form of summer schools and student forums.

J. C. García-Prada and C. Castejón (Eds.): *New Trends in Educational Activity in the Field of Mechanism and Machine Theory*, MMS 64, pp. 91–102, 2019.
https://doi.org/10.1007/978-3-030-00108-7_11

Within the above frames, universities of different IFToMM Member Organizations organize common teaching and exchanges of students and teachers, even in bilateral way within or without well-established programs.

In this paper, an example of such opportunities and, indeed, experiences is reported as gained by the authors from Timisoara (Romania) and Cassino (Italy). The IFToMM spirit is the base of the presented collaboration, even with differences in local activities, that makes possible to form young engineers with a vision towards an international cooperation with common standards.

2 History of Teaching Mechanism Theory

2.1 Evolution at Politehnica University of Timişoara

The Mechanism Theory classes have a long tradition at the Politehnica University of Timişoara. In 1945 at the Politehnica Institute TraianVuia (former name) was founded the Chair of Machine Elements and since 1948 the classes on Mechanism Theory started. The classes on Mechanism Theory were developed historically in accordance with the classes on Machine Elements. The first courses on Mechanism Theory were taught by Paul Sulea (1919–1993). An important development was initiated in 1953, as Francisc Viliam Kovacs started to teach the discipline Mechanism Theory and to develop together with Bernard Horovitz research activities on improvement of tooth geometry of gears and synthesis of linkages in automate devices.

The first published books in the MMS field were "Theory of Mechanisms" by Paul Sulea in 1952 [1] and "Theory of Mechanisms and Machines" by Francisc Viliam Kovacs (1929–2009) and Bernard Horovitz (1921–1970) in 1958 [2], Fig. 1. The first labs were started in 1956 in the newly grounded Mechanism Laboratory in 1955. F.V. Kovacs continued to modernize the content of the courses of Theory of Mechanisms and Machines on the basis of the German and Russian mechanisms schools. Between 1957 and 1962 the Chair of Machine Elements was integrated as a working group in the Chair of Strengths of Materials [3].

a) b)

Fig. 1. Covers of the first books published on Mechanisms and Machines Theory (a) by Paul Sulea [1]; (b) by Francisc Kovacs and Bernard Horovitz [2]

In 1962, the chair was reborn as Chair of Machine Elements and Mechanisms and recruiting young teaching staff members. Young new members were integrated in the teaching staff of the chair in charge to teach Mechanism and Machine Theory, such as Dan Perju (1935–2016), Mihail Crudu (1936–2011), Iaroslav Oprea (1922–2003) and IlieIdiţoiu in 1962, Ştefan Anghel (1941–2003) in 1957, George Gustav Savii in 1973, Ioan Văcărescu (1944–2000) in 1975, Mesaroş-AnghelVoicu in 1978, Valeria Văcărescu in 1980, InocenţiuManiu in 1980 and others.

In 1969, Francisc V. Kovacs, [4], defended his PhD thesis, as later did Dan Perju and Mihail Crudu in 1971. After establishing the ARoTMM as national IFToMM MO in 1972, in Timişoarathe MTM Conference series was initiated with four year frequency. Over 1500 papers and contributions presented at the different sessions of the conference are posted on the site of the digital library dmg-lib.org. The conference MTM became international at its eleventh edition, in 2012 [6].

Francisc V. Kovacs and Dan Perju as representative professors of the chair started the classes of the specializations Precision Mechanics in 1976 and Robotics in 1990. In 2001 the chair of Machine Elements and Mechanisms was converted into the Department of Mechatronics and its new director, InocentiuManiu together with ValerDolga founded the specialization Mechatronics in the same year.

Figure 2 shows the professors of Mechanism Theory, who were also scientific advisors for PhD theses in the field, up to now-days.

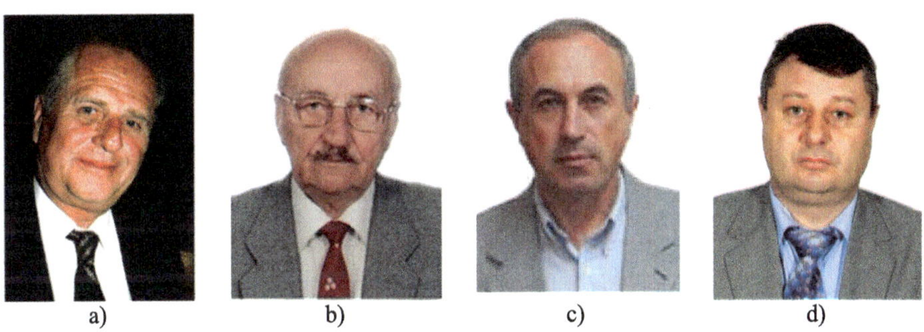

Fig. 2. Professors of Mechanism Theory and PhD advisors in the field of Mechanism Science from UPT. (a) Francisc V. Kovacs (1929–2009); (b) Dan Perju (1935–2016); (c) InocentiuManiu (1952–); (d) Erwin-Christian Lovasz (1967–)

2.2 Evolution at University of Cassino and Southern Lazio

The School of Engineering of Cassino University was started in 1986 and just after, in 1988, a course on TMM was given by Luigi Papa (1944–1997). The course was planned with a classical program as in Italian universities, within about 120 h in a year-long teaching. Textbooks of the Italian tradition were of reference, although the teacher used to provide notes to the students. The TMM team was enlarged, and since 1997, Marco Ceccarelli was the teacher of the main course, since in those years additional teaching was given on Mechanics of robots and Mechanics of vibration.

Textbooks by Marco Ceccarelli were published in English and Spanish, Fig. 3.

a) b)

Fig. 3. Cover of textbooks by Marco Ceccarelli: on Mechanics of robots (a), [5], and on Mechanism design (b), [6]

Later, within the reform of teaching in Italy, the teaching on MMS was planned with a basic course of nine credits at bachelor level, which is nowadays even split in one of six credits for all industrial engineering students and an additional three credits for mechanical engineering only. The previous courses are today programmed in the master curriculum as in the courses of Design of mechanisms, Mechanics of robots, Mechanics of actuators and Mechanics of mechanical systems (given in English).

All the courses are taught with a duration of a semester for nine credits in about 90 h, including exercise time and lab practice if planned.

The teaching of MMS in Cassino is focused on giving the fundamentals of mechanics of rigid bodies for application in designing and operating mechanical systems. Specific teaching is given to design purposes for mechanisms as part of mechanical systems and machinery, but with specific interest on robots and automatic systems.

2.3 Content of the Classes

In Timisoara, the Mechanism Theory classes are scheduled in all programs of study at the Mechanical Engineering Faculty, namely one semester at Mechanical engineering, Transportation engineering, Industrial engineering, Materials engineering, Automotive engineering and two semesters at Mechatronics and Robotics.

The curriculum of the second year in bachelor studies contains the module of Mechanism Theory, with the structure in Table 1.

Table 1. The structure of the module Mechanism Theory in hours per week

Period	Course	Seminar	Laboratory	Project
First semester (14 weeks)	2	1	1	0
Second semester (14 weeks)	2	0	1	1

The course aims to develop students' skills on structural, kinematic, kinetostatic and dynamic analysis of mechanisms, and synthesis of linkages, gear and cam mechanisms. The goal is to acquire knowledge that is required for the development of general engineering applications. The applicative component of the subject is deepened within the laboratory and project activities. The content of the course and seminars is detailed in Table 2 [7, 12].

Table 2. The content of the module Mechanisms Theory

Content of the course (chapters/subchapters)
1. Structure of mechanisms
1.1. Elements and kinematic pairs
1.2. Kinematic Assur groups and connections
1.2. Structural analysis of mechanisms. Applications
2. Kinematic analysis of linkages
2.1. Kinematic analysis by means of the graphical method. Applications
2.2. Kinematic analysis by means of the analytical method. Applications
2.3. Kinematic analysis by means of the graphical - analytical method. Applications
3. Kinematic analysis of gear mechanisms
3.1. Kinematic analysis of ordinary gear mechanisms. Applications
3.2. Kinematic analysis of cycloidal gears mechanisms. Applications
4. Synthesis of gear mechanisms
4.1. Generation of cycloidal and involute profiles
4.2. Geometry of involute spur gears and parallel helical gears
4.3. Elements for gearing check
5. Synthesis of cam mechanisms
5.1. Types of cam mechanisms. Motion curves
5.2. Determination of minimum cam size
5.3. Determination of cam profile
5.4. Kinetostatics and stress analysis of cam mechanisms
6. Synthesis of linkages
6.1. Synthesis of linkages with imposed positions
6.2. Synthesis of path generation linkages
6.3. Synthesis of function generation linkages
7. Kinetostatic and dynamic analysis of mechanisms
7.1. Classification of forces acting upon mechanisms
7.1. Kinetostatic analysis of kinematic groups and connections
7.2. Dynamic motion equations of mechanisms

The teaching means for the courses are the traditional blackboard and/or the modern video projection on screen. For the seminar classes, the students can see the problems on the web page of the department and they can work with specific software applications, such as KOSIM, APPROX, MecaPLAN, SAM, WinDAM, GCAD, KINEMA_5, GIM.

In Cassino the contents of the TMM courses are based on general programs that are adopted at national level, but with specific attention to the specific expertise that the Cassino team has gained in specific topics. Thus, the basic course at bachelor level is given today in two parts, namely a mandatory section of six credits with the fundamentals, and an optional section of three credits with formulation and algorithms for computer-based analysis of planar mechanisms. The fundamentals include the main concepts and basic formulation of mechanics of rigid bodies as applied to planar mechanisms of one DOF. In addition, in both sections information and formulation are given for the main aspects of mechanical transmissions, such as gears, cams, belt chains, bearings, brakes and flywheels. This course is organized with lecture and numerical exercises during the classes.

The master courses are today planned with a considerable activity of practice and lab experiences as a direct application of the lecture topics. In addition, seminars are given on specific topics, even with last advances by teachers, PhD students and visiting professors. In particular, the master course on Design of mechanisms is devoted to formulation and procedures for designing planar mechanisms and mechanical transmissions, by considering also performance evaluation. The master course on Mechanics of actuators is devoted to explain actuators and their operation in machinery by considering dynamic effects and efficiency.

The master course on Mechanics of robots is planned mainly on kinematics and dynamics of robots in both serial and parallel chains, with a consistent part of lab practice with the Adept SCARA robot available at LARM. In addition, the course is planned with practice in numerical evaluation of robot mechanism through simulations that are developed by student teams, using commercial software and their own codes.

The course on Mechanics of mechanical systems is given in English as preferably directed to students in Erasmus program. The lectures are organized to give main concepts and outlines of algorithms for analysis and design of mechanisms that student teams are asked to apply in a detailed study of a mechanical system of their choice, with design issues and performance evaluation.

The interaction of the students with teachers is also an important part of the teaching, thanks also to the fact that in Cassino the master courses are in general attended by no more than 25 students.

2.4 Laboratory of Mechanism Science

In Timisoara, in the Laboratory of Mechanism Science the students perform the following subjects, at the choice of the conductor of classes:

1. Structure of mechanisms
2. Determination of angular speed
3. Kinematic analysis of the universal joint
4. Experimental determination of the motion curves of cams
5. Determination of the critical pressure angle for a prismatic pair
6. Kinematic analysis of linkages and cam mechanisms
7. Mechanical efficiency of a gear box
8. Dynamic balancing of rotors

9. Experimental generation of the involute profile
10. Experimental generation of the cam profile

The laboratory classes are developed in rooms that are devoted only for this activity and are equipped with a large collection of models of all types of mechanisms, experimental test beds (Fig. 4) and a computer network. Examples of mechanism models are shown in Fig. 5.

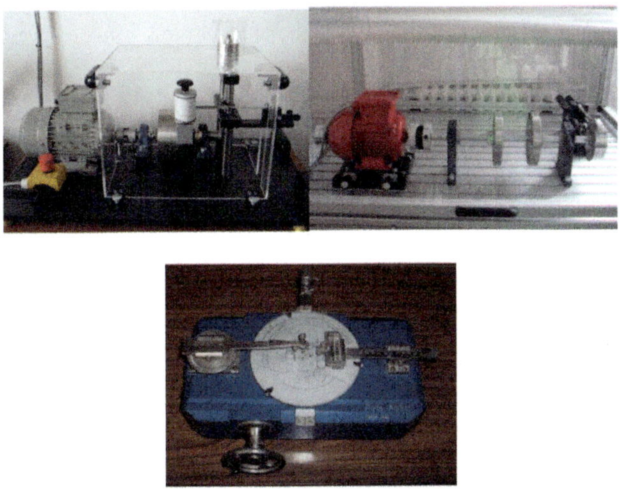

Fig. 4. Different experimental test beds from the Mechanisms Laboratory at UPT

In Cassino, the lab activity supporting teaching is carried out at LARM, the Laboratory of Robotics and Mechatronics, through lab practices and experiences mainly for thesis elaboration. LARM was established in 1990 as since then it has also served as room hosting students and visiting scholars, also for teaching activities [13].

Figure 6 shows the current status of the large room (of about 120 square meters) of the lab with the facilities including desk positions, industrial robots and several prototypes that have been developed through research activities at LARM, also with contributions of students and visiting scholars, for their thesis work.

Examples of those prototypes as mainly due to students work are shown in Fig. 7. In particular, the prototype in Fig. 7c has been designed by students from Timisoara, in Cassino, during an Erasmus period of study.

The lab activities supporting MMS teaching can be summarized in the following aspects:

- Mechanism design (topology search, synthesis, formulation, etc.)
- Operation analysis (dynamic simulation, performance evaluation, actuation analysis, etc.)
- Mechanical design (CAD design, stress analysis, optimization, etc.)
- Prototype construction (components selection, 3D printing of parts, assembly, etc.)
- Testing (lab experiments, prototype tests, experimental validation, etc.).

Fig. 5. Examples of mechanism models from the collection in the Mechanisms Laboratory at UPT

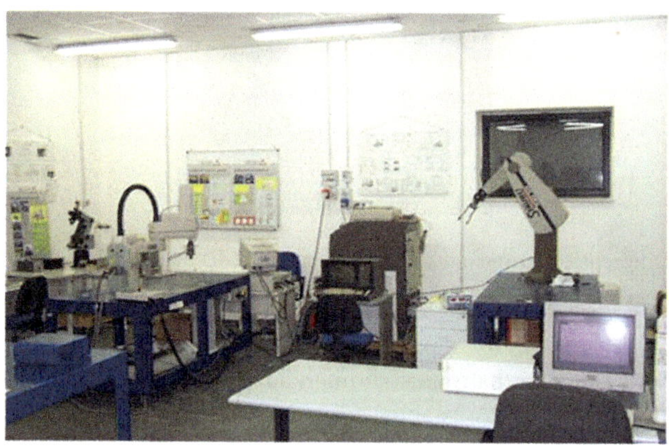

Fig. 6. The room of LARM lab in Cassino

A specific strategy is asked to the visiting students and Cassino students in sharing problems and activities, even if not of their specific interest, with the aim of making experience of team work and coordinating/supervising each other as depending of the skills of each one.

2.5 Student Project Work on Mechanism Science

In Timisoara, the main subjects of the student project work consist of a series of two mechanisms (gears and cams). The chapters of the project refer to:

1. Synthesis of a gear mechanism (two stages)
2. Synthesis of a cam mechanism.

Figure 8 provides a scheme of the components in students' project.

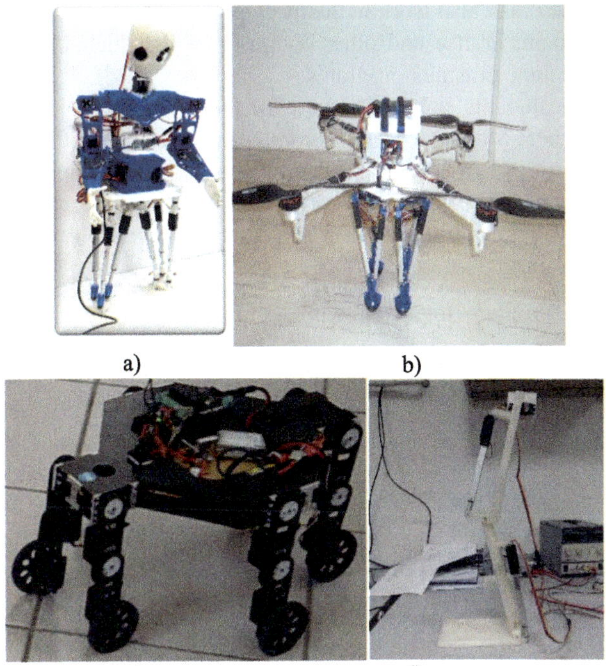

Fig. 7. Examples of prototypes developed at LARM in Cassino with contributions of student projects: (a) LARMbot Humanoid; (b) Heritagebot platform; (c) Cassino Hexapod; (d) leg exoskeleton mechanism

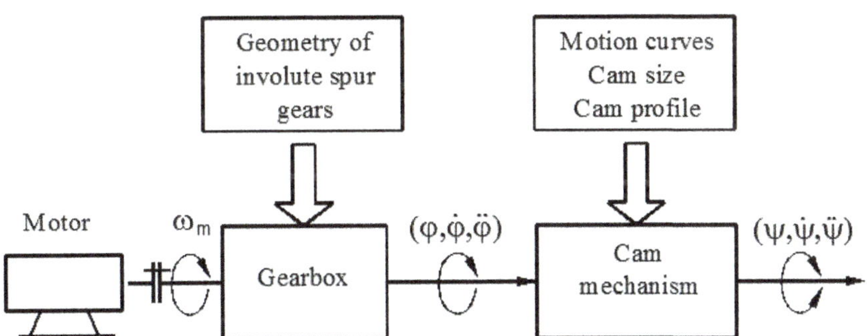

Fig. 8. Student project work at the Mechanism Theory in Timisoara

The students can use the computer network in the Mechanism Laboratory and can perform the computation and design by means of the available software (MathCAD, Excel, AutoCAD).

Students both at the level of bachlor and master programs carry out the work on their degree thesis in the Mechanism Laboratory, if the subjects they chose are related to mechanisms. They may also work in the mechanism laboratories of universities from Germany, Italy, Spain, France and other countries, with which UPT has partnership agreements or common Erasmus programs.

In Cassino, the students at LARM work for their (thesis) project under the direct supervision of teachers Marco Ceccarelli or Giuseppe Carbone with the cooperation of other students (also to learn how to work in a team), by developing, in general, the following aspects:

Acquiring experience with systems and prototypes available at LARM as background for their project

- Describing the problem to attach with identification of requirements and basic formulation
- Developing a solution of their suggestion by characterizing its feasibility with operation analysis and performance evaluation, even with experimental work
- Whenever possible, constructing a prototype by using mainly an approach for low-cost user-oriented solution in order to validate the design results and to check the operation of the solution
- Whenever possible, using the prototype and other equipment available at LARM for experimental validations and practical characterization of the prototype.

As mentioned above, most of the student projects are worked within LARM activities. In addition, the students, either visitor or local ones, are invited to have collaborations with others (students and teachers) either at LARM or with short period of study abroad (mainly within Erasmus program for the Cassino students).

2.6 PhD Theses on the Field of Mechanism Science

In Timisoara, the first PhD advisor in the field of Mechanism Theory was prof.dr.ing. Francisc V. Kovacs, who was accredited to lead PhD students' thesis in 1974. Until his retirement, prof.dr.ing. Francisc Kovacs completed the advisorship of 25 PhD theses in the area of mechanisms and robotics. He was followed by prof.dr.ing. Dan Perju, who became scientific advisor for PhD dissertations in 1990. He led 13 PhD dissertations in the fields of mechanisms and precision mechanics.

Since 1974, the work of research and elaboration of PhD theses related to mechanism theory, under structural, kinematic and dynamic aspects, developed continuously, spanning over 40 years. The achievements can be observed and analyzed in the archive of the university (http://www.upt.ro/Informatii_arhiva—teze-de-doctorat_291_ro.html).

During the last years, the PhD students had the chance to complete or deepen their fundamental and practical knowledge by means of working in the laboratories of other universities such as University of Cassino, Technische Universität Dresden, University of the Basque Country, RWTH Aachen and others, within the framework of Erasmus program or on the basis of bilateral agreements.

In Cassino, the PhD students work their research mainly within LARM research projects and possibly in cooperation with other institutions.

Up to now, more than 20 PhD dissertations have been supervised by Marco Ceccarelli, even with co-supervision from abroad and by Giuseppe Carbone. Since

2010 it have been also possible to plan double PhD degrees with foreign institutions, as an additional aspect of the collaboration. Most of the PhD dissertations were with PhD degree at Cassino University on subjects related both to Mechanism Design and Service Robotics, with theoretical studies for design algorithms and performance evaluation with novel concepts and formulation that have been applied for solution developments up to new prototype constructions and their experimental characterization and validation.

Recently, patent submissions have been also planned with good design results, beside of publication of papers in international journals. More information is available at www.larmlaboratory.net.

3 Current Challenges and Future Developments

Today challenges in MMS teaching can be summarized as coming from the past experience and programs, and from the fast evolving needs of the technology and professional world, in terms of:

- Multidisciplinary formation, requiring to integrate mechanical engineering and particularly MMS with other disciplines in mechatronic approaches, although the fundamentals and cores of mechanics still play a basic role
- Vision of the young engineers open and able to develop new solutions, both in design and operation of mechanical/mechatronic systems
- Skills of using/operating complex new systems with proper efficiency and user-oriented performance
- Capability to understand and identify the lack of information and knowledge that can be solved with proper further formation and collaboration with other professionals, entities, and academic teams, even in programs of continuous education
- Experience in team work sharing problems and solutions, getting also inspiration from the past.

The above challenges indicate main expected developments in the teaching frames as well as in the teaching results.

As recently experienced, reforms and evolutions of teaching as well as the structure and activity of the academic world will solicit and constraint the teaching to a fast evolution to a better oriented formation to a practical implementation of knowledge and technology that MMS can provide, together with more interaction and collaboration of universities from different countries.

4 Conclusions

This paper presents a comparative analysis of activity for formation in MMS fields in Timisoara and Cassino as an example of similarities and differences in MMS teaching in Romania and Italy, even with historical records. It can be noted as MMS is given in both universities as an important area in the formation in any level of engineering education, so that collaboration has been worked out not only within Erasmus program, but even in research projects as an extension of common teaching interests.

References

1. Sulea, P.: Course on Mechanism Theory (Curs de Teoria Mecanismelor). Editura Institutului Politehnic Timisoara, Timisoara (1952). (in Romanian)
2. Kovacs, F., Horowitz, B.: Theory of Mechanisms and Machines (Teoriamecanismelorşimaşinilor), Litografia Institutului Politehnic Timisoara (1958). (in Romanian)
3. Kovacs, V.F., Ionescu, N.: Short history of the Chair of Mechanisms and Machine Elements from the University Politechnica Timisoara (1945–1996) (Schităaistorieicatedrei de Organe de MaşinişiMecanisme de la Universitatea "Politehnica" din Timişoara (1945–1996)). In: Proceedings of the VII MTM and XIII Robotica Symposium 1996, pp. 139–163 (1996). (in Romanian)
4. Lovasz, E.-C., Răduleascu, C., Ciupe, V., Gruescu, C., Maniu, I.: Francisc Viliam Kovacs (1929–2009). In: Ceccarelli, M. (ed.) Distinguished Figures in Mechanism and Machine Science: Their Contibutions and Legacies, History of Mechanism and Machine Science, vol. 26, pp. 159–185. Springer, Dordrecht (2014)
5. Ceccarelli, M.: Fundamentals of Mechanics of Robotic Manipulation. Springer, Dordrecht (2004)
6. Lopez-Cajùn, C.S., Ceccarelli, M.: Mecanismos: Fundamentos cinematicos para el diseno y la optimizacion de la maquinaria, 2nd edn. Trillas, Ciudad de Mexico (2013)
7. Kovacs, F., Crudu, M., Perju, D.: Mechanisms - Course (Mecanisme – curs), Litografia UTT (1992). (in Romanian)
8. Kovacs, F., Perju, D., Văcărescu, I., Mesaroş-Anghel, V., Savii, G., Văcărescu, V.: Synthesis of Mechanisms – Course (Sinteza Mecanismelor – curs), vol. I, II, Litografia UTT (1992). (in Romanian)
9. Lovasz, E.C., Cărăbaş, I.: Principles of Synthesis of Gear and Cam Mechanisms (Principii de sinteză a mecanismelor cu roţi dinţate şi came). Editura Politehnica, Timişoara (2006). (in Romanian)
10. Luck, K., Modler, K.-H.: Getriebetechnik - Analyse, Synthese, Optimierung, 2nd edn. Springer, Berlin (1995)
11. Manolescu, N., Kovacs, F., Oranescu, A.: Mechanisms and Machines Theory (Teoria mecanismelor şi a maşinilor). Editura Didactică şi Pedagogică, Bucureşti (1972). (in Romanian)
12. Perju, D.: Mechanisms for Precision Mechanics (Mecanisme de mecanică fină), Litografia UPT, vol. 1 + 2, Timişoara (1990). (in Romanian)
13. Ceccarelli, M.: An illustrated history of LARM in Cassino. In: Proceedings of RAAD 2012 International Workshop on Robotics in Ale-Adria-Danube Region, Napoli, Edizioni Scientifiche e Artistiche, pp. 35–42 (2012)

Using an Android App for Teaching Mechanism and Machine Theory

Eduardo Corral Abad(✉), María Jesús Gómez García, Ramona Ruiz Blázquez, Alejandro Bustos-Caballero, and Juan Carlos García-Prada

Universidad Carlos III de Madrid, Madrid, Spain
ecorral@ing.uc3m.es

Abstract. In recent years, the results obtained by the Engineering students of the subject Mechanism and Machine Theory at the Universidad Carlos III of Madrid are not the expected results. This is due to the lack compression of the students to the theory of the subject. This forces teachers to find new ways to help students to reach the desired level. So, the present work focuses in helping students to enhance their knowledge about the theoretical concepts of Mechanisms and Machines. Following the idea of promoting the student's personal work and the idea that the student does not have too much time, the designed tool will be an Android App tool. The Android App has been programmed with Android Studio. It is intended to achieve two goals, firstly, it is accessible and, secondly, it is portable.

Keywords: Android · App · E-learning · Accessible · Portable Teaching

1 Introduction

The aim of the subject "Mechanism and Machine Theory" is not only to provide students with theoretical knowledge, but also to enhance their compression of it. For this reason, the subject must be divided into theory, problems, laboratories and symposiums.

This division and method of learning is correct when the student devotes the right time. But, because in the regular assessment the student does not have much time, the student tries to learn to solve the problems without understanding the concepts. This forces teachers to find new ways to help students to reach the desired level.

For this reason, we are searching for training methods based on new principles and using computer technologies [1]. Nowadays, the "Mechanism and Machine Theory" has already been updated with E-learning [2], but the problem that the E-learning does not solve is the lack of time. We were more focused on help the student with a tool that can be used at any place at any time [3]. It is possible thanks to Android. Every student has an own smartphone or tablet. That give us the opportunity to teach them by this way.

Nowadays, many new methods are used for e-learning like the knowledge pills or the virtual labs [4]. However, although the pills are widely used for theoretical

J. C. García-Prada and C. Castejón (Eds.): *New Trends in Educational Activity in the Field of Mechanism and Machine Theory*, MMS 64, pp. 103–110, 2019.
https://doi.org/10.1007/978-3-030-00108-7_12

explanations, at Mechanical Engineering studies the explanation of basic exercises and the theoretical concepts provides a new and more difficult scene. Teaching Mechanism and Machine Theory is a good example of these cases.

Knowledge pills are a tool that has been used recently obtaining good results [5]. But the biggest disadvantage that they have is the necessity of a pc or laptop, and the time that they require. As we said before, we were more focused on helping the student with a tool that can be used at any place at any time without the necessity of having much time.

Following the idea of promoting the student's personal work, the designed tool is an interactive and portable tool. The proposed application has been programmed with Android Studio, and is presented as an executable file accessible from smartphones and tablets. It allows the student to interact and check the theoretical concepts, and it provides the level of knowledge the student?

2 Objective of the App

The main aim of this new technique is to pass on from the reproductive cognitive activity to the search activity and intends to use modern technologies and new training methods which are based on new principles and techniques that are closely associated with the computer technologies.

We would like to improve the theoretical understanding of "Mechanism and Machine Theory" in the students.

We thought that the best way is not only with theory (just memorizing). It is needed that the students think, understand, and also memorize. For this reason, it is needed some App that gives theory concepts, checks if they know or not, and also, explains and shows the concepts to learn. It is possible to achieve all this goals using tests.

The App is designed using interactive test. They can check if the students know the concept. If they don't know, the test shows the correct answer and some explanation if it is needed. Also, the test is a great tool for the students to check the level of understanding. Sometimes, the students think that they have already understood all, and nevertheless, they do not understand all.

The tests can show easily theoretical concepts that the students didn't understand or know. Also, it is not only memorizing and copying the theory without understanding, the student must understand the question and the theory and think in order to find the correct answer.

Other point that we wanted is that the tests are not so long. This is because we wanted the possibility that the student could check fast if he is ready, or what concepts are not not understood.

And the last objective, is the portability. We manage this thank to Android and the smartphones and tablet. It is important, that the student can do this test fast and without the necessity of a computer (with computer and time the student can use the E-learning). The student can do the test in the way of the library to know where he should start to study, or just in some free time that he is waiting, and check if he needs more time to study or if he is ready for the exam.

3 Operating System: Android

The smart mobile phones are changing people's life in a new way. The new devices offer capabilities similar to a personal computer with cameras, processors, GPS receivers or accelerometers. Without forgetting, that the main advantage of these smart phones is that they are small and you can carry them in your pocket.

This revolution has started in the last years, more specifically in 2005, when Google bought Android INC. Android INC was a small company that designed applications for smart phones. Two years later, the Open-Handset-Alliance was created in 2007, and then, the first version of the Android software was delivered in 2008. Since then, android has continued growing and expanding to the present [6–9].

Nowadays, the operating system of Android is the most used method of creating interactive clients using Java language [10]. 85% of the world market share uses Android, above other platforms such as iPhone or Windows Phone. In Spain, 90% use Android [6], reason enough for being the chosen system for this project.

Also, it is well known that programming in Android offers the possibility of being able to access a wide range of devices [7].

As mentioned above, Android is still the main software for Smart phone. This is mainly because it is open source and free source. Android software also has many other advantages, such as: powerful API (application programming interface), API libraries and uses JAVA structure. Thanks to this, it can be used in all types of CPUs and has an acceptable level of security [6].

4 Software Architecture

Android is an open-source software stack that includes the operating system, middleware, and key mobile applications, along with a set of API libraries for writing applications that can shape the look, feel, and function of the devices on which they run. There is even a full application framework included, so third-party apps can be built and installed, and all applications have equal standing. Native and third-party applications are written with the same APIs and executed on the same run time [8].

The Android stack or Android OS architecture, in Fig. 1, that consists of several dependent parts or layers. The upper layers base their operation on the lower layers. The applications that we develop are executed at the top of the stack, in the Applications layer. These applications make use of multiple managers that Android provides in its Application Framework layer, including the window manager and location manager, databases, telephony, and sensors. In turns they all need open-source libraries for application development more easily, such as SQLite, WebKit, OpenGL, and are made in native C code. Next to the libraries, the Android Runtime is composed by the Dalvik Virtual Machine (DVM) and the core libraries that provide Android specific functionality. The DVM is responsible for translating the bytecode of the applications into native code understandable by the device. Finally, the Linux-based kernel is the core of the Android operating system, which provides a low-level interface with the hardware, memory management, and process control, all optimized for mobile and embedded devices [8, 11].

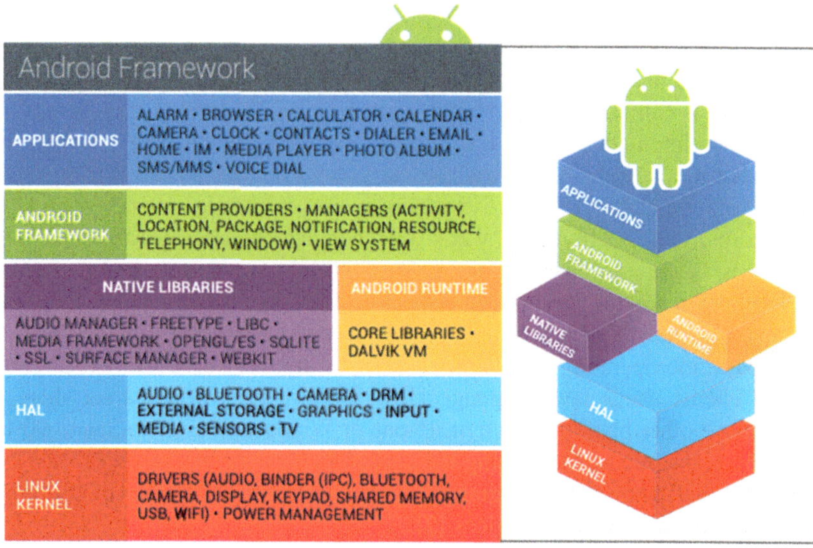

Fig. 1. Android Stack (https://source.android.com/)

One of Androids key architectural goals is to allow applications to interact with one another and reuse components from one another. This reuse applies not only to services, but also to data and the user interface [9].

When an app is compiled, the files generated are.dex (Dalvik Executables), that is, the result of compiling the generated java code and combining it with the libraries used in.dex files. The part of the Android system that executes or runs the compiled DEX code is the DVM. The DVM itself is a piece of software written in another language that runs on a specially adapted version of the Linux operating system. The compiled Java code, along with the other resources, is placed in a bundle of files called an Android application package (APK), and this is what the DVM needs to run the app [12].

5 Project Structure. Android Studio

The project has been made with Android Studio, the official IDE (Integrated Development Environment) created by Google for Android apps development, shown in the Fig. 2.

Android applications have some files that are required and some that are optional. Figure 3 shows the structure of this android project. First of all, the AndroidManifest. xml is the most important file of the app, it is the Android application descriptor file where the name, the contents and behaviour of the application are defined, and it lists all application's activities and services, along with the permissions and features the application needs to run. The java file is a folder containing all of the source code of the application, and the res file is the folder containing the resources like drawable, layout,

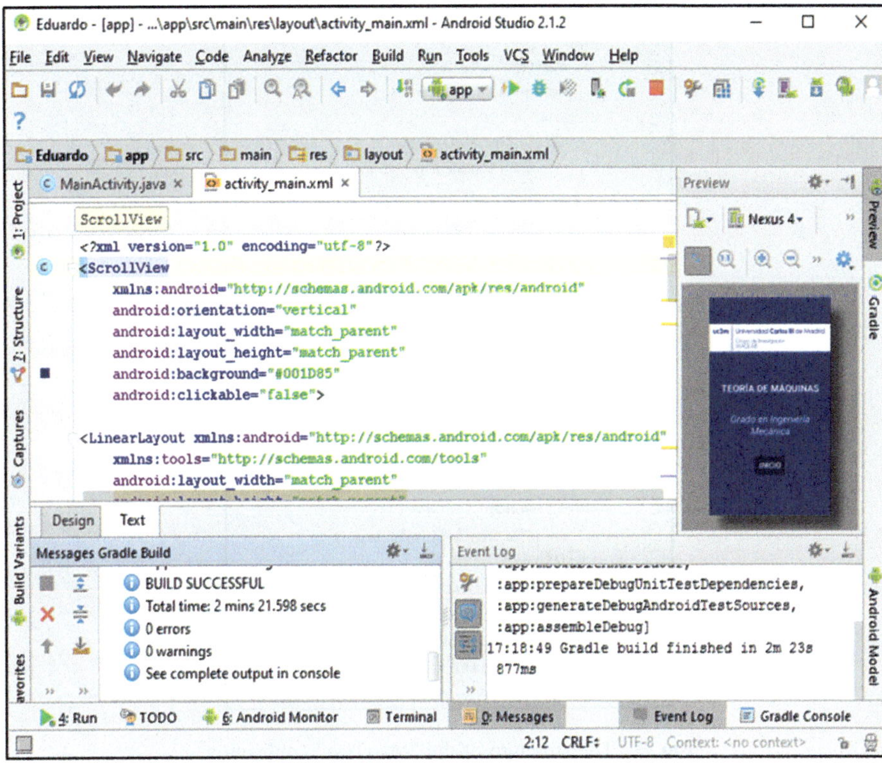

Fig. 2. Android Studio windows

mipmap and values folders. The drawable folder contains the images used by the application. In the layout folder are the XML files that create the views of the app. The mipmap folder contains the icon image to launch the application in different sizes for the different screen dimensions. And in the values folder there are other resources used by the application such as styles and colors, and the strings file which contains all the text used in the tests. Finally, Gradle is the build system.

In summary, to design and develop Android apps, all components in Android are a piece of code that has a well-defined life cycle. To make an interactive Android program, it must start by sub-classing the Activity class. Activities provide the reusable, interchangeable parts of the flow of UI (user interface) components across Android application. An Android activity is both a unit of user interaction and a unit of execution [10]. Each screen of the app will be defined by an activity that consists of two files, a.java file that describes the logic part and an.xml file that describes the graphic part of the screen.

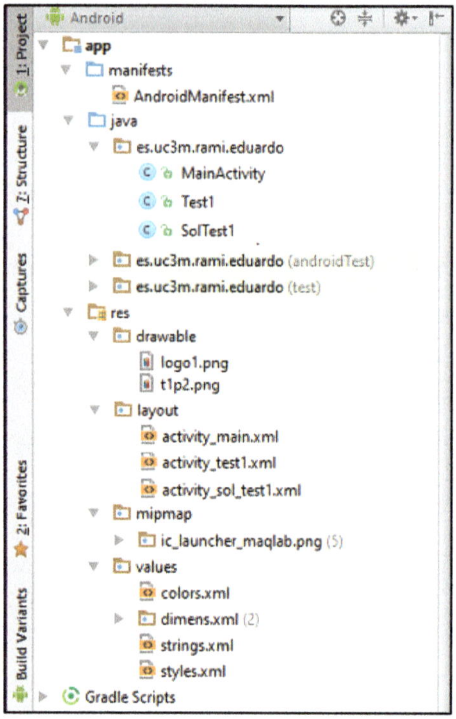

Fig. 3. The structure of the "MaqTest" application

6 App Algorithm

The algorithm of the iteration tests is shown in Fig. 4. This is the algorithm or scheme of the window/activities of the app. the user can access to the first window/activity (Main Activity) and then they can go to the second window/activity (start the App). After this, the user will go to the next window/activity (i-Test) and choose the desired test.

In the i-test Activity, the user selects the answer and later, the user could check the answer or see the correct answer. Then, the user could go to the next test or come back to the second window/activity (test Activity) and start a new test.

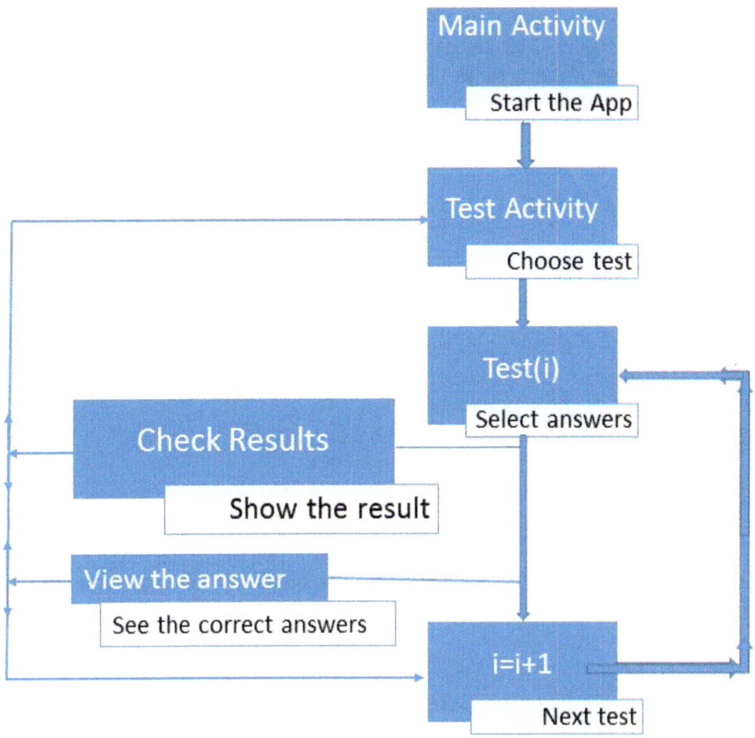

Fig. 4. The App algorithm

7 Conclusions

A new an interested tool for the learning of the students has been presented. The authors believe that it is an accessible and portable tool to help students to understand the theory of mechanism and machine theory.

The results of the students have been improved.

The first response of the students has been very satisfactory.

The authors believe that it is a complementary method to the translated methods and the e-learning.

In the current times where the student does not have enough time, with this application, we offer the students an easier way to study and understand concepts anywhere.

It is also possible to conclude that in order to quickly evaluate the student's deficiencies the best way is a test.

Another important point to realize is that Android is free software, and it will not cost anything to the students. Besides, they don't need to buy any new computer or to have a good internet connection, once the app is installed.

In the future we will create more Apps for more subjects: for design, calculations, and for English and Erasmus students.

Acknowledgments. The authors wish to thank the Spanish Government for financing provided through the MCYT project "RETOS2015: Sistema de monitorización integral de conjuntos mecánicos críticos para la mejora del mantenimiento en el transporte-maqstatus" and the MINECO Grant BES-2014-070462.

References

1. Suñer, J.L., Mata, V.: Utilización de programas de simulación en la docencia de la ingeniería mecánica. In: Actas del X Congreso de Innovación Educativa en las Enseñanzas Técnicas, Valencia, Spain (2002)
2. Chaturvedi, S.K., Dharwadkar, K.A.: Simulation and visualization enhanced engineering education development and implementation of virtual experiments in a laboratory course—American Society for Engineering Education, AC 2011-742 (2011)
3. Fu, T.: Applications of computer simulation in mechanism teaching. Comput. Appl. Eng. Educ. **11**, 156–165 (2003)
4. Mateo Sanguino, T.J., Andújar Márquez, J.M.: Simulation tool for teaching and learning 3D kinematics workspaces of serial robotic arms with up to 5-DOF. Comput. Appl. Eng. Educ. **20**, 750–761 (2012)
5. Nickerson, J.V., Corter, J.E., Esche, S.K., Chassapis, C.: A model for evaluating the effectiveness of remote engineering laboratories and simulation in education. Comput. Educ. **49**, 708–725 (2007)
6. Gironés, J.T.: El gran libro de Android. Marcombo, Barcelona (2012)
7. Aliferi, C.: Android Programming Cookbook. Exelixis Media P.C., Attica (2016)
8. Meier, R.: Professional Android™ 4 Application Development. Wiley, New York (2012)
9. Komatineni, S., MacLean, D.: Pro Android 4. Apress, New York (2012)
10. Mednieks, Z., Dornin, L., Meike, G.B., Nakamura, M.: Programming Android. O'Reilly, Newton (2011)
11. Lequerica, J.R.: Desarrollo de aplicaciones para Android. Anaya, Madrid (2013)
12. Horton, J.: Android Programming for Beginners. Packt Publishing Ltd., Birmingham (2015)

Dynamics of Adaptive Mechanisms - New Division of TMM

K. S. Ivanov(✉)

Almaty University of Power Engineering and Telecommunication,
Almaty, Kazakhstan
ivanovgreek@mail.ru

Abstract. The existing course of dynamics contains dynamic research of mechanisms with one degree of freedom and with one input. Only such mechanisms are used in mechanical engineering. However, recently essentially new so-called adaptive mechanisms - gear variators have been developed. The adaptive gear variator represents the one-drive wheelwork with two degree of freedom which is independently adapting for variable force loading by change of motion speed. Action of an adaptive variator is described by dynamics. In the paper the dynamics of transitive modes of the adaptive mechanism with two degrees of freedom is investigated. The received results are a basis of essentially new course of dynamics named «Dynamics of adaptive mechanisms».

Keywords: Gear variator · Power adaptation
Dynamics of adaptive mechanisms

1 Introduction

The existing course of dynamics contains dynamic research of mechanisms with one degree of freedom and with one entry. Only such mechanisms are used in engineering industry. However, recently brand new so-called adaptive mechanisms - gear variators have been developed. The gear variator represents the mechanism with two degree of freedom. The main difference of a gear variator from all mechanisms existing in the world: the gear variator has only one entry. It means that on a way of actuating the gear variator is the usual one-drive mechanism. But on behaviour of act the gear variator implements a discovery - «Effect of force adaptation in the mechanics».

The adaptive gear variator represents a wheelwork with constant engagement of the toothed wheels and having ability to be adapted for variable force loading at the expense of independent (without a control system) change of motion speed.

Attempts to create a gear variator (the adaptive gearing) were undertaken by many inventors [1–4]. In the basis of Ivanov's invention the two-mobile planetary kinematic chain [3, 4] was used. It has been proved [5–7] that if the kinematic chain contains the mobile closed contour at traffic with two degree of freedom then the closed contour imposes additional communication on traffic of links and provides definability of traffic. At the same time the kinematic chain gets property of power adaptation to a variable load. Such property takes place in traffic operating condition at a relative uniform motion of all links.

J. C. García-Prada and C. Castejón (Eds.): *New Trends in Educational Activity in the Field of Mechanism and Machine Theory*, MMS 64, pp. 111–122, 2019.
https://doi.org/10.1007/978-3-030-00108-7_13

However on the start (in the motion beginning) the output link is motionless, and the kinematic chain has one degree of freedom. Transfer cannot to transfer force to a target link for a beginning of motion. Definability of motion is absent. The using of brake on one of mobile links [2] demands a control and deprives autonomy transfer. Use of dynamic inertia parameters on the start [3] provides a small starting moment and is not reliable.

Reliable start with overcoming of the high starting moment can be provided by creation of the dead center position of the closed contour at chosen sizes of links [8]. Starting communication is created by choosing of matching sizes of links and is eliminated after start. For elimination of starting communication after the beginning of motion of all links the additional transfer which wedges out the closed contour is used [8]. This additional transfer takes place in parallel with the basic two-mobile planetary kinematic chain and does not obstacle to its motion. The wedging out of additional transfer can be created on the basis of coincidence of linear speeds of some links [9].

The developed kinematic analysis and force analysis of an adaptive variator in a uniform motion with two degree of freedom [10–12] install analytical regularity of interconnection of the kinematic and force parameters according to mechanics laws. Numerical instances prove the found regularity.

However autonomy of motion on start at transition from a condition with one degree of freedom into a two-mobile condition, and also in operating condition at loading change should be analyzed. It is necessary to describe and investigate dynamic transient for an estimation of reliability of autonomy and an efficiency of gear variator.

Work is devoted to description and research of dynamics of transients of adaptive gear variator.

The received results are a basis of essentially new course of dynamics named «Dynamics of adaptive mechanisms».

2 Description of Adaptive Gear Variator

At the description of the device and work of the adaptive gear variator we will use following designations:

M_9, M_{10} – external moments on input 9 and output 10 carriers,
F – input impellent,
R – output force of resistance,
r_9, r_{10} – radiuses of input 9 and output 10 carriers,
u_{9-5}^{pl} – transfer ratio of a planetary kinematic chain from the input carrier 9 to the output satellite 5,
u_{9-5}^{ad} – transfer ratio of additional transfer from the input carrier 9 to the output satellite 5,
z_i $i = 1, 2, 3, \ldots 8$ – numbers of teeth of wheels,
ω_9, ω_{10} – angular velocities of input 9 and output 10 carriers.

The adaptive gear variator (Fig. 1) contains following details: input carrier 9, input satellite 2, block of solar wheels 1–4 fixed on intermediate shaft, block of ring wheels 3–6 leaning against satellites, output satellite 5 and output carrier 10.

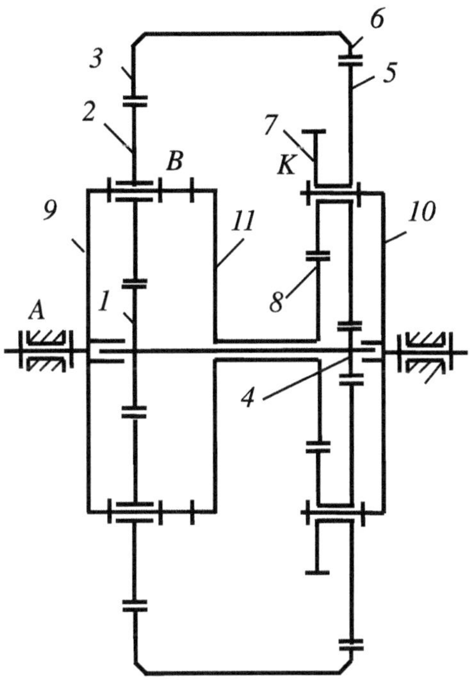

Fig. 1. Adaptive gear variator

Input 9 and output 10 carriers are executed with equal sizes (radiuses) $r_9 = r_{10}$, that matches to the formula connecting numbers of teeth of wheels 1, 2 and 4, 5 $z_1 + z_2 = z_4 + z_5$. It leads to transfer wedging on start.

Additional transfer is executed in the form of the gearing from the input carrier 9 to the output satellite 5 containing a toothed wheel 8 connected to the input carrier 9 by means of disk 11 and a wheel 7 rigidly connected to the output satellite 8. Additional transfer provides the wedging out of the kinematic chain after the motion beginning. Additional transfer in the form of wheels 8 and 7 has transfer ratio u_{9-5}^{ad} and doubles a planetary kinematic chain from input carrier 9 to output satellite 5 with transfer ratio u_{9-5}^{pl}. The equality $u_{9-5}^{ad} = u_{9-5}^{pl}$ takes place, where

$$u_{9-5}^{ad} = -z_8/z_7, \tag{1}$$

$$u_{9-5}^{pl} = \frac{u_{13}^{(9)} - u_{46}^{(10)}}{u_{56}^{(10)}(u_{13}^{(9)} - 1)}. \tag{2}$$

Here $u_{13}^{(9)} = -z_3/z_1$ - the transfer ratio of wheels 1 and 3 at the motionless carrier 9, $u_{46}^{(10)} = -z_6/z_4$ - the transfer ratio of wheels 4 and 6 at the motionless carrier 10, $u_{56}^{(10)} = z_6/z_5$ - the transfer ratio of wheels 5 and 6 at the motionless carrier 10.

After substitution of these values in (2) we will gain $u_{9-5}^{n\pi} = \frac{z_3z_4z_5 - z_1z_5z_6}{z_3z_4z_6 + z_1z_4z_5}$.

From a condition of equality of the transfer ratios expressed by formulas (1) and (2), we will gain a condition of interconnection of numbers of teeth of wheels of the mechanism allowing synthesizing the mechanism

$$\frac{-z_8}{z_7} = \frac{z_3z_4z_5 - z_1z_5z_6}{z_3z_4z_6 + z_1z_4z_5}. \tag{3}$$

The adaptive gearing works as follows.

In the motion beginning (at start) the output carrier 10 is motionless, transfer has one degree of freedom and can be in free move at relative mobility of wheels 1–4, 2, 3–6, 5 of closed contour. Force interacting of links of transfer is presented on a side view (Fig. 2). The load 11 with weight G and mass m, attached by a flexible filament to shaft A of output carrier 10, creates a tractive resistance. Relative motion of links on the start is possible generally when carriers 9 and 10 have different radiuses, and transfer has eccentricity $e = r_9 - r_{10}$, allowing creating moment $M = Fe$ turning the satellite 5 around motionless point K of the output carrier 10.

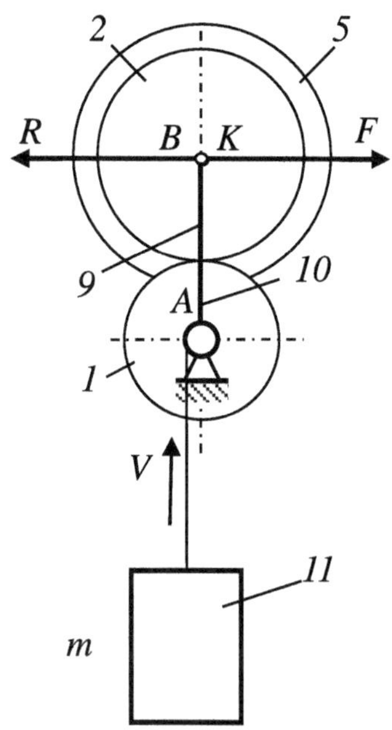

Fig. 2. An adaptive gear variator on a side view

Thus closed contour from toothed wheels gets internal relative mobility. However at equal radiuses of carriers 9 and 10 the kinematic chain of transfer appears wedged because the line of acting of impellent F from of the party of the input carrier 9 in point B passes through point K of the output carrier 10 and force F is directed oppositely to resistance force R. Eccentricity is equal $e = 0$, and the driving moment which rotates the output satellite 5 and all closed contour in relative motion is absent. As a result of wedging the kinematic chain loses one degree of freedom and can begin motion only in the chocked condition, overcoming force of resistance R and an output starting moment of resistance on the carrier 10. Start from a place becomes absolutely reliable (as in the usual mechanism with one degree of freedom).

After start from a place additional (parallel) transfer (Fig. 1) through wheels 8 and 7 provides transfer of the driving moment directly from the input carrier 9 on the output satellite 5, providing its relative motion in the closed contour, and eliminates wedging. The mechanism passes into a condition with two degree of freedom with relative mobility of links of a contour. In this condition the equilibrium of the mechanism is carried out by a principle of possible works with adaptation to a variable output moment of resistance by formula which is resulted in [6]

$$\omega_{10} = \frac{M_9 \omega_9}{M_{10}}. \tag{4}$$

Thus, the offered design provides automatic overcoming of high starting resistance and definability of motion in two mobile condition.

3 Dynamics of Transient of Gear Variator in Stage of Running Start

For the dynamic analysis we will use the theorem about change of kinetic energy: change of a kinetic energy of the mechanism within some interval of a time is equal to work of external forces.

$$A_M - A_R = T - T_0. \tag{5}$$

T, T_0 - kinetic energy in end and in beginning of motion interval,
A_M, A_R - work of an impellent and work of force of resistance.

Conveniently instead of works A_M, A_R to use powers of an impellent and force of resistance P_M, P_R. For this purpose we will divide the Eq. (5) into an interval of time t, we will gain

$$P_M - P_R = (T - T_0)/t. \tag{6}$$

On initial transient two kinds of transitive motion take place: (1) Start - overcoming of the starting moment of resistance during time of start t_s, (2) Acceleration - before transition into an operational mode of uniform motion during time t_a. General time of initial transient $T_b = t_s + t_a$.

Let's determine parameters of the start. Kinetic energy of the mechanism can be determined under formula $T = 0.5mV^2$, where m - the reduced mass of all links, V - speed of a point of reduction. However for decision simplification it is possible to neglect masses of links of the mechanism as they are small in comparison with mass of moving load. Then $T_0 = 0$, $T = 0.5mV_s^2$, where m - mass of a load (Fig. 2), V_s - the given starting initial speed of motion of load V_s, $V_s = \omega_{10s}r_{10}$ - starting initial angular speed, $V_s = \omega_{10s}r_{10}$ - output shaft radius.

Power of resistance on start $P_{Rs} = GV_s$, where G - a load weight ($G = mg$).

From the formula (6) we will determine a required power of the engine for start providing.

Let's determine parameters of the acceleration after the start. Interconnection of parameters of acceleration in matching with formula (6) has following format

$$P_M - P_R = (T_a - T_s)/t_a, \tag{7}$$

where t_a - time of acceleration.

From here

$$\begin{aligned} P_M &= P_{Rs} + 0.5m(V_a^2 - V_s^2)/t_a = P_{Rs} + 0.5m(V_a - V_s)(V_a + V_s)/t_a \\ &= P_{Rs} + 0.5ma_a(V_a + V_s)/t_a \text{ or} \\ P_M &= P_{Rs} + 0.5ma_s(V_a + V_s), \end{aligned} \tag{8}$$

where $a_a = (V_a - V_s)/t_a$ - given (allowable) initial acceleration.

On driving power it is possible to select the engine (propeller) with matching angular velocity ω_M and to determine the driving motion moment

$$M_M = \frac{P_M}{\omega_M}. \tag{9}$$

The driving moment must be checked on a condition of serviceability of transfer [8] using radiuses of toothed wheels 1, 4, 9, 10 and the given moment of resistance $M_{Rs} = P_{Rs}/\omega_{10s}$

$$M_9 \geq M_{Rs}\frac{R_4R_9}{R_1R_{10}}. \tag{10}$$

After that it is possible to determine the start time under the formula received from the Eq. (7)

$$t_s = \frac{mV_s^2}{2(P_M - P_{Rs})}. \tag{11}$$

After the start the acceleration begins - the sped up motion with transition into motion operating condition. The Eq. (6) becomes

$$P_M - P_{Rs} = 0.5m(V^2 - V_s^2)/t_a, \tag{12}$$

where t_a - acceleration time, V - the given speed of motion of a load in the end of initial transient. The acceleration proceeds before achievement of equality of powers of impellents and forces of resistance. We will determine acceleration time

$$t_a < \frac{0.5m(V^2 - V_s^2)}{P_M - P_{Rs}}. \tag{13}$$

Further the motion operating condition begins. At equality of powers of an impellent and force of resistance ($P_R = P_M$) the uniform motion without kinetic energy change occurs. At instant change of a moment of resistance the transitive regime with change of a kinetic energy before achievement of equality of powers of impellents and forces of resistance at the expense of change of output angular speed according to the formula (6) occurs. In this case driving power P_M remains without change, power of resistance matches to the changed moment resistance M_R at former angular velocity ω_{R0}, that is $P_R = M_R\omega_{R0}$, and the kinetic energy in the end of transient will match to new angular speed $\omega_R = P_M/M_R$. Then from the formula (6) it will be possible to determine time t of transient to the changed moment of resistance in motion operating condition. The further motion will be uniform.

4 Numerical Instance of Dynamic Calculation

Let's execute dynamic calculation of the mechanism (Figs. 1 and 2).

Initial data:

Load weight and mass of load $G = 10000\ Nm,\ m = 1000\ Ns^2/m$.

The kinematic parameters of the beginning of motion

$$V_s = 0.1\ m/s,\ V_a = 1\ m/s,\ a_s = 7.27\ m/s^2.$$

Geometrics $R_1 = 0.044,\ R_4 = 0.012,\ R_9 = R_{10} = 0.056,\ r_{10} = 0.010$.

To determine: a time of start t_s, acceleration time t_a, a time of the beginning of motion T_b for transition into motion operating condition with $V = V_a = 1m/s\ V = V_a = 1m/s$ and with operating condition motion parameters $M_R,\ \omega_R$. Numerical results we will show on the diagram of a tractive characteristic of a gear variator (Fig. 3) for visualization.

The solution

1. Output angular velocity on start $\omega_{10s} = V_s/r_{10} = 0.1/0.01 = 10s^{-1}$.
2. A output moment of resistance on start $M_{Rs} - Gr_{10} = 10000 \cdot 0.01 = 100Nm$.
3. Power of resistance on start $P_{Rs} = M_{Rs}\omega_{10s} = 100 \cdot 10 = 1000Nm/s$. Points A and B.

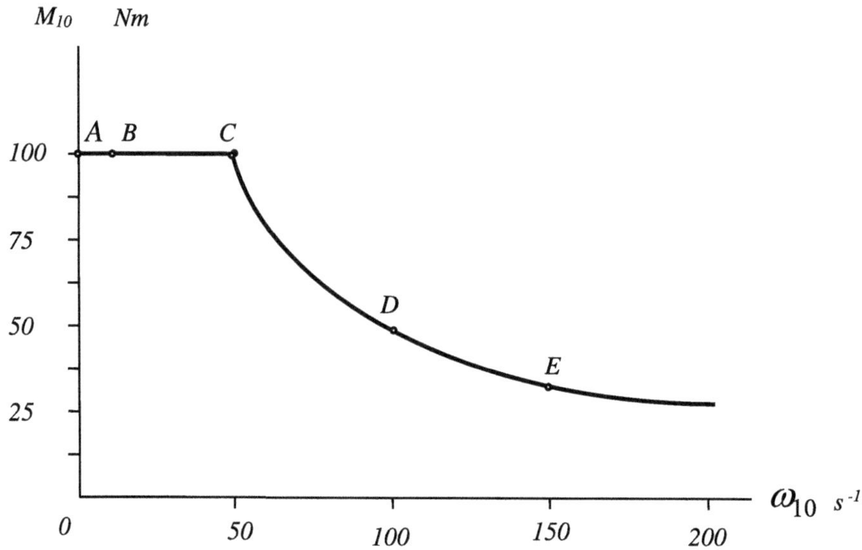

Fig. 3. Tractive characteristic of gear adaptive variator

4. The driving power consumed for overcoming of starting resistance and acceleration on the Eq. (8)

$$P_M = P_{Rs} + 0.5ma_s(V_a + V_s) = 1000 + 0.5 \cdot 1000 \cdot 7.27 \cdot (1 + 0.1) = 5000Nm/s.$$

   We select the electric motor: power of 5 kW, a rotational speed of 1500 rpm ($\omega_M = \omega_9 = 150\ s^{-1}$).
5. Driving moment $M_M = M_9 = \frac{P_M}{\omega_M} = \frac{5000}{150} = 33.3Nm$.
   Check of the driving moment on possibility of start $M_9 \geq M_{Rs}\frac{R_4 R_9}{R_1 R_{10}} = 100 \frac{0.012 \cdot 0.056}{0.044 \cdot 0.056} = 27.2Nm$.
6. A start time (getaway) by formula (11)

$$t_s = \frac{mV_s^2}{2(P_M - P_{Rs})} = \frac{1000 \cdot 0.1^2}{2(5000 - 1000)} = 0.00125\ s.$$

   Start is presented by section *AB*.
7. Acceleration time after start

$$t_a = \frac{m(V_a^2 - V_s^2)}{2(P_M - P_{Rs})} = \frac{1000 \cdot (1^2 - 0.1^2)}{2(5000 - 1000)} = 0.123\ s.$$

   Dispersal is presented by piece *BC*.
8. A time of the beginning of motion - transition from start into motion operating condition $T_b = t_s + t_a$.

5 Dynamics of Transient of Gear Variator in Stage of Steady Motion

After the beginning of motion in operating condition with a constant resistance moment the uniform motion with constant angular velocity takes place. Power of resistance is equal in the beginning of operating condition to driving power $P_R = P_M = 5000Nm$. Parameters of power of resistance are equal $M_R = 100Nm$, $\omega_{10} = 50s^{-1}$ - the point C.

In motion operating condition at a process with decrease of a moment of resistance there is an increase of angular velocity and kinetic energy increase according to the formula (6). For example, with decrease of a moment of resistance to meaning $M_R = 50Nm$ we will gain

$$\omega_{10} = \frac{P_M}{M_R} = \frac{5000}{50} = 100s^{-1} - \text{ point } D.$$

In this case in a transitive regime on formula (13) it is necessary to use following parameters.

In point C initial speed is equal $V_s = \omega_{10} r_{10} = 50 \cdot 0.01 = 0.5m/s$, in point D matching to new amount of moment of resistance $M_R = 50Nm$, the terminal speed is matching to angular velocity $\omega_{10} = 100\ s^{-1}$ and is equal $V = \omega_{10} r_{10} = 100 \cdot 0.01 = 1m/s$. Driving power P_M is remaining invariable. Power of resistance is equal

$$M_R = 50Nm.$$

Then by formula (13) we will gain time of acceleration and of transition into a new regime of motion with changed moment of resistance M_R and with power P_R

$$t_a = \frac{0.5m(V^2 - V_s^2)}{P_M - P_R} = \frac{0.5 \cdot 1000 \cdot (1 - 0.5^2)}{5000 - 2500} = 0.15s.$$

In the end of this transitive regime (point D) input and output powers will be made equal again, and there will be further a regime of a uniform motion with parameters of point D.

Generally transitive process in operating condition occurs at change (decrease or increase) a moment of resistance and power of resistance that leads to respective alteration of speeds of motion of links and kinetic energy of a mechanism.

When the minimum moment of resistance which equals to driving moment $M_R = 33.3Nm$ takes place we will gain $\omega_R = \frac{P_M}{M_R} = \frac{5000}{33.3} = 150\ s^{-1}$ - point E.

6 Numerical Instance of Dynamic Calculation

Initial data match to the previous instance.

Power of resistance is equal in the beginning of operating conditions to driving power $P_R = P_M = 5000Nm$. Parameters of power of resistance are equal in the beginning of a stage of installed motion $M_{Rb} = 100Nm,\ \omega_{10b} = 50s^{-1}$ - point C. New value of resistance moment $M_{Rn} = 50Nm$.

To determine the output angular velocity ω_{10e} in the stage of the installed motion operating condition at end of motion interval and transition time in a new regime of motion t.

Solution

1. According to the formula (6) final angular velocity $\omega_{10e} = \frac{P_M}{M_{Rn}} = \frac{5000}{50} = 100\ s^{-1}$ - in point D. An initial winding speed of load in point C $V_b = \omega_{10b} r_{10} = 50 \cdot 0.01 = 0.5m/s$. Speed in end of an interval $V_e = \omega_{10e} r_{10} = 100 \cdot 0.01 = 1m/s$.
2. Power of resistance in the beginning of an considered interval of motion - in point C $P_{Rn} = M_{Rn}\omega_{10b} = 50 \cdot 50 = 2500Nm/s$.
3. Transition time in a new regime of motion

$$t = \frac{m(V_e^2 - V_b^2)}{2(P_M - P_{Rn})} = \frac{1000 \cdot (1^2 - 0.5^2)}{2(5000 - 2500)} = 0.15\ s.$$

4. At the new minimum moment of resistance equal to driving moment $M_{Rn} = 33.3\ Nm$, we will gain $\omega_{10n} = \frac{P_M}{M_{Rn}} = \frac{5000}{33.3} = 150\ s^{-1}$ - point E. The speed of load $V_e = \omega_{10n} r_{10} = 150 \cdot 0.01 = 1.5\ m/s$.
5. Transition time in a regime of motion from position in point C with the minimum resistance - point E.

$$t = \frac{m(V_e^2 - V_b^2)}{2(P_M - P_{Rn})} = \frac{1000 \cdot (1.5^2 - 0.5^2)}{2(5000 - 2500)} = 0.4\ s.$$

Parameters of tractive characteristic: Point A - start with parameters $\omega_{10} = 0,\ M_R = 100\ Nm$. AB - transitive motion regime on start with overcoming of starting resistance (parameters of point B:$\omega_{10} = 10\ s^{-1},\ M_R = 100\ Nm$). BC - a section of a regime of acceleration and of increase of kinetic energy. Point C - the beginning of operating condition of motion with two degree of freedom (parameters $\omega_{10} = 50\ s^{-1},\ M_R = 100\ Nm$). CD - transitive regime of operational motion in a condition with the reduced moment of resistance (angular velocity and a kinetic energy are increasing). Point D - an intermediate state of operating condition of motion (parameters $\omega_{10} = 100\ s^{-1}\ M_R = 50\ Nm$). The return motion (for example, from point D to point E) takes place when moment of resistance is increasing with decreasing of angular velocity and a kinetic energy.

7 Conclusion

The gear variator is created on the basis of a kinematic chain with two degrees of freedom that determines its basic difference from existing transfer mechanisms. Research of dynamics of an adaptive gear variator allows presenting a full picture of its action in all regimes of motion. The elementary method based on the theorem of change of kinetic energy is used for dynamic research.

The start transitive regime of motion provides the start from place and acceleration of mechanism. The start takes place when the motion power is exceeding power of force of resistance. Reliability of the beginning of motion provides the starting wedging of a kinematic chain. The gear variator overcomes the maximum starting resistance in the accelerated motion with increase in kinetic energy.

Operational motion regime (after breaking) occurs at equality of motion power and power of resistance force. Motion is uniform. At change of balance of power the corresponding transitive regime changing parameters of motion and kinetic energy takes place. Here the closed contour of the mechanism gets the compelled internal mobility and provides adaptation to variable loading.

The made dynamic analysis confirms the efficiency and reliability of work of the two-mobile mechanism containing necessary additional constraints, in all regimes of motion.

In the paper the dynamics of transitive modes of the adaptive mechanism with two degrees of freedom is investigated. The received results are a basis of essentially new course of dynamics named «Dynamics of adaptive mechanisms».

References

1. Crockett, S.J.: Shiftless, continuously-aligning transmission. Patent of USA 4,932,928, Cl. F16H 47/08, U.S. Cl. 475/51; 475/47.1990, 9 p.
2. John, H.: Power transmission system comprising two sets of epicyclical gears. Patent of Great Britain GB2238090 (A) (1991). 11 p.
3. Ivanov, K.S., Yaroslavtseva, E.K.: Way of automatic and continuous change of the torque and speed of rotation of output shaft depending on resistance to movement and the device for its realisation. Patent of Russia RU № 2398989, 09 October 2010. 10 p.
4. Ivanov, K.S., Almaty, K.A.Z.: Owner of the registered sample. The name - Device of automatic and continuous change of rotation moment and change of rotation speed of output shaft depending on resistance moment. The deed on registration of the registered sample № 20 2012 101 273.1. Day of Registration 02.05.2012. The German patent and firm establishment. Federal Republic Germany (2012). 12 p.
5. Ivanov, K.S.: The question of the synthesis of mechanical automatic variable speed drives. In: Proceedings of the Ninth World Congress on the Theory of Machines and Mechanisms, vol. 1, August 29th–September 2nd, 1995. Politechnico di Milano, Italy, pp. 580–584 (1995)
6. Ivanov, K.S.: Discovery of the force adaptation effect. In: Proceedings of the 11th World Congress in Mechanism and Machine Science, vol. 2, Tianjin, China, 1–4 April, 2004, pp. 581–585
7. Ivanov, K.S., Tultaev, B.: Toothed continuously variable transmission (CVT)–industrial realization. In: New Trends in Mechanism and Machine Science. Theory and Applications in Engineering, Transactions of 4-th Euro Conference on Mechanisms, Santander, Spain, 18–22 September 2012, vol. 7, pp. 329–335. Springer (2012)
8. Ivanov, K.S.: Proof of existence of a gear variator as wheelwork with constant engagement of toothed wheels. In: Proceedings of the Third Conference MeTrApp 2017, Trabson, Turkey, pp. 83–90 (2017)
9. Ivanov, K.S.: Creation of adaptive-mechanical continuously variable transmission. In: 5th International Conference on Advanced Design and Manufacture (ADM 2013), Valencia, Spain, pp. 63–70 (2013)

10. Ivanov, K.S.: The simplest automatic transfer box. In: WCE 2010, World Congress on Engineering 2010, (ICME), London, UK, pp. 1179–1184 (2010)
11. Ivanov, K.S.: Paradox of mechanics–a basis of creation CVT. In: Transactions of 2-d IFToMM Asian Conference on Mechanisms and Machines Science, Tokyo, Japan, 7–10 November 2012, pp 245–264 (2012)
12. Ivanov, K.S.: Theory of continuously variable transmission (CVT) with two degrees of freedom. Paradox of mechanics. In: Proceedings of the American Society of Engineers Mechanics (ASME) International Mechanical Engineering Congress and Exposition (IMECE 2012), Houston, Texas, USA, pp. 543–562 (2012)

Generated Graphics and Game Development Software in Engineering Education: Perspectives and Experience of Usage

Gleb Procurat(✉), Alexandra Strukova, Andrei Vukolov, and Olga Egorova

Robotics and Complex Automation Faculty, Institute of Modern Educational Technologies, Bauman Moscow State Technical University, Moscow, Russian Federation
dvil200220@yandex.ru, twdragon@bmstu.ru

Abstract. Engineering education requires modern training courses and integrated IT solutions. This paper describes the development of 3D demonstration utility which imitate involute gear wheel manufacturing process. Development of such utilities now is significantly simplified by special software called "3D engines". Students and TMM teachers now have perspectives to easily develop their own utilities and models within the training according to famous "Russian method" of engineering education. The example and experience of development of such solution is given in this paper.

Keywords: Software · 3D graphics · TMM · Engineering education Course development · Unity · Game engine · Virtual reality

1 Introduction

TMM training requires from teacher to use very large set of demonstrative examples from times of creation of the discipline [20]. In modern era of graphics, video games, 3D solutions and augmented reality [11] many attempts made to transfer such examples, models and works into virtual environment. As an example of such work the 3D model of Watt's mechanism described in [22] can be considered. It consists of 3D model (Fig. 1*b*) rendered to precisely conform the original mechanism (Fig. 1). The 3D model realized as motile assembly including parts connected between each other through constraints which are imitating kinematical pairs.

Original model designed in SolidWorks but can be easily transferred to specialized software like ANSYS and SolveSpace. In virtual environment even the task of demonstration of assembling and disassembling can be solved. But there is the only main disadvantage that can eliminate all advantages described above: necessity of original software suite to perform operations and even to

J. C. García-Prada and C. Castejón (Eds.): *New Trends in Educational Activity in the Field of Mechanism and Machine Theory*, MMS 64, pp. 123–130, 2019.
https://doi.org/10.1007/978-3-030-00108-7_14

a

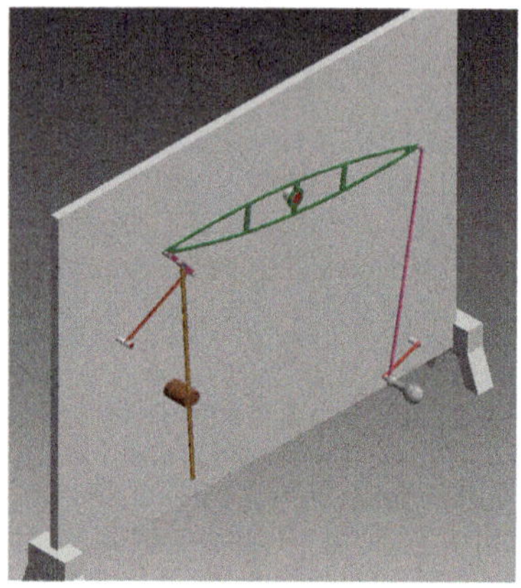

b

Fig. 1. Watt's mechanism

demonstrate of 3D model. It is not possible to bundle the work into the standalone demonstration suite or to perform parametrization of parts. Also it is completely impossible to bring into model the ability to change configuration or characteristics of model in real time. Thus usage of the model in the training process is very limited.

The opportunity to do all of the described things is to generate models by predefined procedures directly in runtime. This is the usual technique in game development [14,15,24] and cinematic 3D modeling. Content of 3D scene or level fillings are usually being generated in real time [18]. Also the projects of training-oriented games are known [5] which generate full user interface and environment to demonstrate problems such as machine learning. In CAD systems and finite-element analysis systems the procedural mesh generation used for solid bodies modeling [17]. Most scientific solutions use constant predefined models to generate geometrical objects [12]. Only CAM solutions use physical model of deformations to regenerate 3D models.

Possibility to create model which could be changed in runtime can be realized using special library kits integrated with pipelines called "3D engines" [9]. The "engine" includes unified model data transport and tensor processing unit (TPU) which performs geometrical transformations. Rasterization layer connects geometrical models and transformation matrices with operating system environment and graphic driver. It also realizes the standard set of algorithms for game development and model-to-model interactions. These peculiarities make it possible to easily integrate 3D engines into the other projects like robotics simulation [4]. Only licensing constraints limit the opportunities [10,16,21] but in this case the open source solutions become significant for education [6]. Below the description of usage of 3D game development engine in realization of TMM training task will be given.

2 Target Setting

The TMM course in Bauman University includes a set of practice works. One work dedicated to involute gearing exploration [19]. This work includes drawing of involute gear wheel contour divided by sectors with different values of instrument's displacement value. The task on this practice work is to build a trace of instrument's positions on the paper circle placed on the special device (Fig. 2)[1] which imitates bending method of tooth cutting. The toothed raker moves connected with the circle to provide bending motion. The obtained model represents the state of involute gearing according to displacement. In 3D setting this device demonstrates deformation of the disk under influence of the cutting forces provided by raker moving up and down.

3D model of the considered experiment is simple and easily can be used during training but it requires imitation of cutting. In case of usage of ordinary modeling approach this is necessary to use CAM solution which contains physical model integrated.

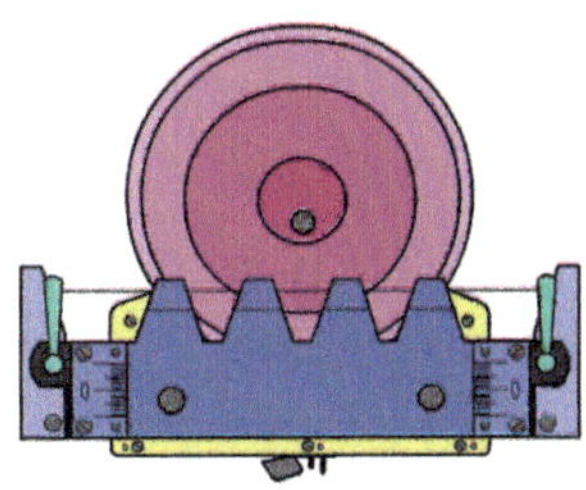

Fig. 2. Gear wheel modeling device (sketch)

2.1 Selection of the Development Kit

To avoid necessity of high computational resources required for CAM solution and usage of proprietary software the decision was made to use game development engine under conditions of free or non-commercial license [13,21]. Unity 3D was chosen because of following factors:

- Support of free base models produced by free 3D sculpting solution;
- Support of scripting on common programming languages useful for students;
- Royalty-free end-user license, free for non-commercial usage [21];
- Support of scriptable motion, collisions, built-in destroying and partitioning procedures for the model generator;
- Existence of successfully developed educational projects [8,23];
- Possibility to bundle obtained 3D solution into the autonomous software suite available on several operating platforms including Windows and Linux.

Unity 3D engine allows to realize imitation of working process of the device (Fig. 2) using a simple C# script.

[1] Sketch by A. Strukova.

3 Modeling of involute gear wheel creation in Unity 3D

In Unity 3D the 3D object is not the solid body. Instead of solid bodies it uses wireframe meshes constraining the area of space which the 3D object occupies. The following listing in C# shows initialization of the mesh for basic disk-shaped wheel model. The basic model must be prepared and loaded into the engine from any modeling or sculpting solution.

```
public class Make : MonoBehaviour {
    public Vector3 d; //
    public Mesh mesh;
    public float l = 0.5f;
    public GameObject disk;
    void Start()
    {
        d = Vector3.up / -1000;
        mesh = GetComponent<MeshFilter>().mesh;
    }
```

The raker model designed for the following set of the gear wheel parameters:

$$z = 22 \qquad h_a^* = 1$$

$$d = 180\,\text{mm} \qquad \alpha = 20^\circ$$

$$c^* = 0.25 \qquad \beta = 0$$

3.1 Cutting

In fact the cutting of material can be realized in 3D workspace as deletion (boolean subtraction) of set of vertices which represents boolean intersection of models' meshes. Unity 3D does not have internal realization of boolean subtraction algorithm for set of vertices [8]. This peculiarity requires developer to manually create the new set of 3D vertices within the wheel model's mesh. Unity 3D provides a mechanism for running of such procedures called "collisions". This is the typical solution used in game development [2] The collision mechanism tracks intersections between the projections of mesh's lines pairs connected with different models, and then it calls the procedure (collision handler) specified by developer.

Imitation of cutting in this work is handled by the following handler:

```
void OnCollisionEnter(Collision collision)
{
    bool id = false;
    Vector3[] ver = mesh.vertices;
    for (int i = 0; i < ver.Length; i++)
    {
        for (int j = 0; j < collision.contacts.Length; j++)
        {
            Vector3 point = transform.InverseTransformPoint(collision.
                contacts[j].point);
            float dist = Vector3.Distance(point, ver[i]);
            if (dist < 0.0006f)
            {
                ver[i] += d;
                id = true;
            }
        }
    }
```

```
    if(id)
    {
        mesh.vertices = ver;
        mesh.RecalculateNormals();
        mesh.RecalculateBounds();
        GetComponent<MeshCollider>().sharedMesh = mesh;
    }
}
```

For each edge within the involved mesh the collision handler calculates reactive force. It transforms into the value of `Collision.RelativeVelocity` property to obtain displacement matrix for each vertex in the collision area. The collision area is now shaped according to geometry of the raker. The `RecalculateNormals()` and `RecalculateBounds()` functions use the obtained set of vertices caught in the collision area to regenerate mesh of the disk. Unity 3D does not allow user to destroy objects in runtime. Because of that the excluded vertices could only be displaced to center of the disk model and to disappear from rasterization conveyor.

3.2 Movement

Rotation of the wheel and contemporaneous motion of the raker are realized as described in [19]. Each stage of raker movement (Fig. 3*a*) produces a collision which runs the procedure similar to the one described above. The collision handler also runs stepping movement of the raker and linked rotation of the wheel. All of these movements require special algorithms to be realized. In Unity 3D environment these algorithms are easy enough to be understand and written as scripts by stugent. For example, stepping rotation of the wheel performed using transformation matrices called *quaternions* [1]. The quaternion object represents rotation as renormalization of connected coordinate system of the model. However, it requires from developer to deeply understand vectors and normals control. The handling procedure of stepping rotation looks like following:

```
if(b)
{
    if(a < 0)
    {
        transform.position += Vector3.up / -50;
        a--;
    }
    else b = false;
}
else
{
    if(a < 100)
    {
        transform.position += Vector3.up / 50;
        a++;
    }
    else b = true;
}
if(a == 100)
{
    Quaternion e = Quaternion.AngleAxis(8.55f, new Vector3(0,0,1));
    Quaternion g = Quaternion.AngleAxis(-7f, new Vector3(0,1,0));
    disk.transform.rotation *= e;
    raker.transform.rotation *= g;
}
```

Interaction between trapezoidal teeth of the raker and moving wheel generates involute curve of the wheel's teeth as described in [3,19]. This process completely corresponds with definition of the involute teeth manufacturing based on bending method. The same handling procedure was used to test the modeling of gear wheel with 14 internal teeth (Fig. 3*b*). Wide usage of the developed sources makes it possible to add parametrization and code generation features in future [4].

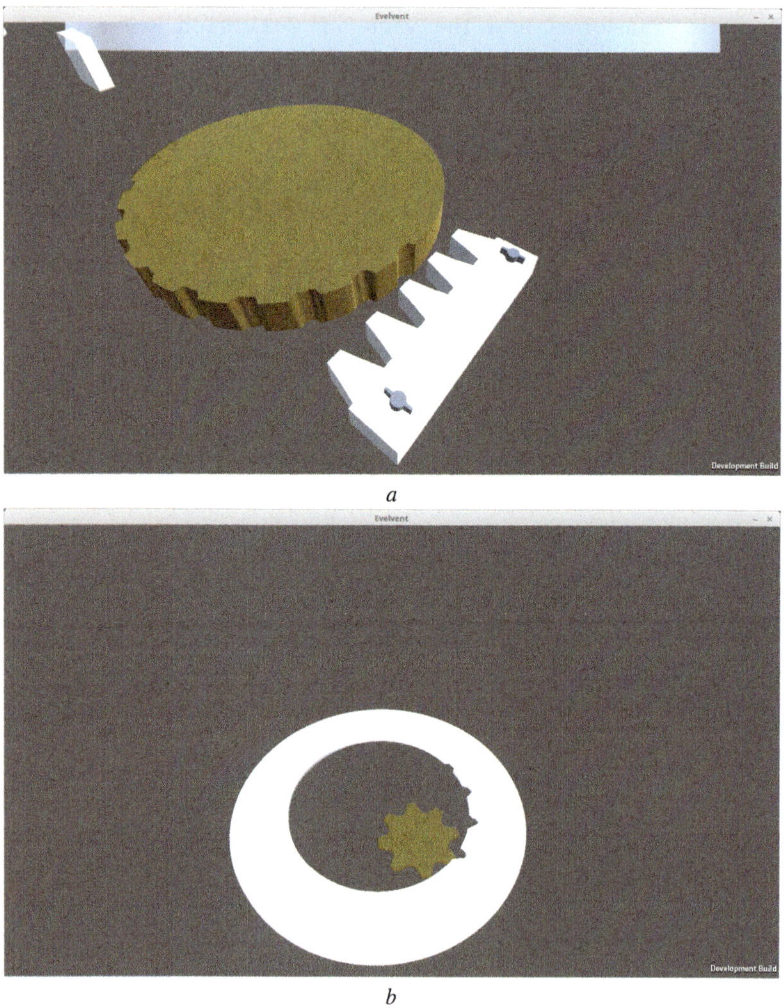

Fig. 3. Imitation of the cutting

4 Conclusion

The developed 3D graphic solution demonstrates not only imitation of manufacturing of involute gear wheel but also the high potential and perspectives of generated graphics within engineering education process. The simple solution available for development by students inside single year realizes huge amount of features including 3D simulation, parametrization and adaptation. The development process greatly improves understanding of physical problems encapsulated within mechanisms and machines because the student writes handling procedures for the collisions himself. High adaptation possibilities allows to use "Russian method" [7] of engineering education in the virtual environment, including augmented reality [11] because the generated models could be easily made with form of the real parts.

Also usage of game development engine extends functionality of ordinary 3D models (including parametrized ones) developed in CAD software with possibility of testing the developed mechanism in completely determined virtual world defined by engine's TPU. For TMM teachers the generated graphics also provides perspectives of development fully interactive laboratories filled with object available to interact and to change. The described opportunities open the way to complement the typical laboratory of mechanics or robotics with elements of virtual reality with realistic quality which modern game development can provide.

References

1. Basmanov, D.: The captious quaternions (2013). https://habrahabr.ru/post/183908/. (in Russian)
2. Berger, F., Müller, W.: Towards an Open Source Game Engine for Teaching and Research, pp. 69–76. Springer, Heidelberg (2012). https://doi.org/10.1007/978-3-642-31439-1_7
3. Colbourne, J.R.: Tooth Profile of an Involute Gear, pp. 24–52. Springer, New York (1987). https://doi.org/10.1007/978-1-4612-4764-7_3
4. Cubero, S.N., Billingsley, J.: Game Development Tools for Simulating Robots and Creating Interactive Learning Experiences, pp. 113–134. Springer, Heidelberg (2015). https://doi.org/10.1007/978-3-662-45514-2_11
5. Derwich, S., Essalmi, F.: A 3D Learning Game for Representing Artificial Intelligence Problems, pp. 33–38. Springer, Singapore (2017)
6. Duran, J., Villagrasa, S.: Teaching 3D Arts Using Game Engines for Engineering and Architecture, pp. 113–121. Springer, Heidelberg (2013). https://doi.org/10.1007/978-3-642-39420-1_13
7. Egorova, O.V.: About "Russian Method" of training of engineers. In: International Affairs, vol. 1 (2014). http://interaffairs.ru/read.php?item=10459. (in Russian)
8. Falah, J., Charissis, V., Khan, S., Chan, W., Alfalah, S.F.M., Harrison, D.K.: Development and Evaluation of Virtual Reality Medical Training System for Anatomy Education, pp. 369–383. Springer, Cham (2015). https://doi.org/10.1007/978-3-319-14654-6_23
9. Freiknecht, J., Geiger, C., Drochtert, D., Effelsberg, W., Dörner, R.: Game Engines, pp. 127–159. Springer, Cham (2016). https://doi.org/10.1007/978-3-319-40612-1_6
10. Ghosh, R., Glott, R., Krieger, B., Robles, G.: Free/Libre and open source software: survey and study. Technical report. International Institute of Infonomics, University of Maastricht (2002)

11. Gu, J., Duh, H.B.L.: Mobile Augmented Reality Game Engine, pp. 99–122. Springer, New York (2011). https://doi.org/10.1007/978-1-4614-0064-6_4
12. Kanai, S.: Content-based 3D mesh model retrieval from hand-written sketch. Int. J. Interact. Des. Manuf. (IJIDeM) **2**(2), 87–98 (2008). https://doi.org/10.1007/s12008-008-0038-4
13. Kovalenko, V.A.: State of deployment analysis of free software in institutions of the educational system in Russia. Pedagog. Educ. Russ. **6**, 188–192 (2013). (in Russian)
14. Lin, Q., Zhao, Z., Xu, D., Wang, R.: Design and Implementation of an OpenGL Based 3D First Person Shooting Game, pp. 50–61. Springer, Heidelberg (2011). https://doi.org/10.1007/978-3-642-18452-9_3
15. Martin, A., Lim, A., Colton, S., Browne, C.: Evolving 3D Buildings for the Prototype Video Game Subversion, pp. 111–120. Springer, Heidelberg (2010). https://doi.org/10.1007/978-3-642-12239-2_12
16. Pires, P.S.M., Rogers, D.A.: Free/open source software: an alternative for engineering students. In: Proceedings of 32th ASEE/IEEE Frontiers in Education Conference (2002)
17. Quadros, W.R., Shimada, K., Owen, S.J.: Skeleton-based computational method for the generation of a 3D finite element mesh sizing function. Eng. Comput. **20**(3), 249–264 (2004). https://doi.org/10.1007/s00366-004-0292-4
18. Ripamonti, L.A., Mannalà, M., Gadia, D., Maggiorini, D.: Procedural content generation for platformers: designing and testing FUN PLEdGE. Multimed. Tools Appl. **76**(4), 5001–5050 (2017)
19. Tarabarin, V.B., Kuzenkov, V.V., Fursyak, F.I.: Laboratory Practice on Theory of Mechanisms and Machines. Bauman Moscow State Technical University, Moscow (2002). (in Russian)
20. Vukolov, A.: F. Reuleaux, F. Wittenbauer: Their Influence on Evolution of Applied Mechanics in Russia at the Beginnings of XXth Century, pp. 315–322. Springer, Cham (2017). https://doi.org/10.1007/978-3-319-44156-6_32
21. Vukolov, A.: Free and Open Source Software Applications for Education of TMM Discipline in Bauman University, pp. 253–260. Springer, Cham (2017). https://doi.org/10.1007/978-3-319-44156-6_26
22. Vukolov, A.Y., Egorova, O.V.: New challenges of 3D mechanisms models in TMM training. Int. Rev. Mech. Eng. **10**(5), 303–311 (2016). https://doi.org/10.15866/ireme.v10i5.9246
23. Wang, D.: Gamified learning through Unity 3D in visualizing environments. Neural Comput. Appl. 1–6 (2017)
24. Warren, R., Champion, E.: Linked Open Data Driven Game Generation, pp. 358–373. Springer, Cham (2014). https://doi.org/10.1007/978-3-319-11915-1_23

MMS in the Engineer Program: New Trends

Tool^{2}Task – A Software Tool for Automatic Task Generation in Mechanism Theory

Jascha Paris(✉), Mathias Hüsing, and Burkhard Corves

Institute of Mechanism Theory, Machine Dynamics and Robotics
RWTH Aachen University, Steinbachstraße 53b, 52074 Aachen, Germany
{Paris,Huesing,Corves}@igmr.rwth-aachen.de

Abstract. A central element of most courses in mechanical engineering are exercises. In small group exercises and self-examination exercises, students can use, practice, and deepen what they have learned. In order to assist the students in this learning process, teachers try to provide a reasonable number of exercises. Against the background of the digitalization of teaching and blended learning concepts such as electronic exams, serious gamification and individual learning paths it is necessary to provide the ability to generate exercises automatically.

Tough, the main challenge is to generate genuinely random tasks. In this contribution Tool^{2}Task, a tool for the automated generation of mechanism theory tasks, is presented. A first step towards individual random tasks was done by developing appropriate algorithms and basic functions for a web based application. This application is able to generate random tasks based on lists of equations, application-related task topics and algorithms for solving the desired types of exercises.

1 Mechanism Theory at RWTH Aachen University

Against the background of the ISEMMS conference, this section will initially provide an insight into the organization of the mechanism theory lectures at RWTH Aachen University. The motivation for automatic creation of exercises is described, also with regard to the didactics, in Sect. 2. Subsequently, the state of the art is partly shown in Sect. 3. The presented web based application Tool^{2}Task is described by mean of three sections. First, requirements and the concept of Tool^{2}Task are set out in Sect. 4. Secondly, a rough overview over the implemented algorithms is given in Sect. 5. Lastly, the implementation and used technologies are described in Sect. 6.

At RWTH Aachen University, mechanism theory is being taught in two courses [1]. The lecture Electromechanical Drive Technology, German abbreviation EMAT, is part of the bachelor's degree in mechanical engineering with specialization in development and design. This course deals mainly with the synthesis and analysis of linkages and cam mechanisms. Different methods for pose

J. C. García-Prada and C. Castejón (Eds.): *New Trends in Educational Activity in the Field of Mechanism and Machine Theory*, MMS 64, pp. 133–140, 2019.
https://doi.org/10.1007/978-3-030-00108-7_15

analysis, pose synthesis, dead point synthesis, velocity analysis and acceleration analysis are treated. The last part of the lecture deals with the dimensioning and selection of electric drives. The master's degree lecture Motion Technology, German abbreviation BWT, is based on EMAT and deals with multiple generation of coupler curves, planetary gears, curvature analysis of trajectories, kinetostatic analysis and compliant mechanisms. In the following, this paper focusses on EMAT, but most of it also applies to BWT.

The EMAT course is attended by 160 students and is worth 5 credit points. Overall, EMAT consists of the educational offerings in Table 1. In lectures the professor explains the theory and derives the methods used. Besides that, a central element of most courses in mechanical engineering are the exercises. Research assistants explain the application of the theory in demonstrational exercises by solving selected tasks. Furthermore, the students have the possibility to attend small group exercises and apply the methods within the geometry application GeoGebra and CAD systems. In order to understand and deepen what they have learned, students should complete the provided tasks and old exams. However, the only must-attend event is the final written exam itself.

Table 1. Educational offerings within the course EMAT

Description	Hours per semester	Type	Lecturer
Lecture	21	Weekly	Professor
Demonstrational exercise	21	Weekly	Research Assistant
Small groups exercise	21	Weekly (opt.)	Student Assistant
CAD exercise	16	Block course (opt.)	Research Assistant
Consultation hours	–	On request (opt.)	Research Assistant
Exercise tasks	–	Self-learning	–
Exam	2	Exam	–

2 Motivation for Automated Task Generation

As already mentioned, exercises are a central element of most courses in mechanical engineering. In small group exercises, self-examination exercises or tutorials students have the ability to use, understand and deepen what they have learned. Topic-specific exercises have an important role in teaching, because they stimulate the students' intense debate with the content of course [2]. A purely receptive way of learning does not lead to the desired deeper understanding of the lecture material, especially in subjects with complex mathematical connections. Above all independent learning exercises play a central role because they give the learner feedback on the learning method used. In order to support students in this learning process, lecturers try to provide a reasonable amount of exercises. However, the creation of such tasks and related solutions is time consuming and therefore limited by resources of the institutes.

If a student has particular difficulties with a specific task type, it is possible that too few tasks are available to him. Especially repeaters, students who have failed in the first attempt, already know most of the provided tasks. Therefore, they are lacking new exercises to prepare themselves for a second exam. In most cases the challenge of exercises lies in finding the right approach to solve the task. Hence, it is desirable to provide many tasks, so that the students can practice themselves in finding the right approaches and decide which types of tasks they need for training.

The introduction of electronic tests and serious gamification even requires the creation individual exercises. Only if each student receives his own tasks, it is possible prevent dissemination of solutions and provide a fair exam and avoid frustration in serious games. In addition, individual exercises provide further advantages. If a student is helped by a fellow, the solution cannot be passed on, instead the students have to talk about the solution, as each student has a different assignment.

The described initial situation also applies to the course EMAT, for which roughly 160 students have enrolled. Although for EMAT exist 77 classical exercises, 15 exercises for geometry software and 5 example exams, for some types of exercises only few tasks are available. Furthermore, the authors' institute wants to build blended learning skills, since digital teaching concepts will gain importance in the future.

In summary, the following motivations should be noted:

- **Individual exercises:** Each student can set his focus himself. Help among students focusses on the methods not solutions
- **Electronic self-tests:** Control of the learning level by means of electronic self-tests
- **Competence in blended learning:** Build blended learning skills for future teaching concepts
- **Serious gamification:** Enable fair and motivative serious games with random and automatically generated tasks
- **Promoting repeaters:** The difficulty is often the approach. Repeaters therefore need more exercises for the second experiment

3 State of the Art of Automated Task Generation

The effort to automatically generate exercises for students or pupils is widely spread because of its benefits and advantages. Though, the variation of those tasks if often limited to pure numbers while the structure of exercises remains unchanged. For example Fikar et al. generate random numbers for their exercises in control engineering [3]. More extensive are commercial applications like for example KUTA Software [4]. This application creates worksheets for maths lessons, generates illustrations and equations. But only one step is required for the solution - so these are trivial problems. Prados et al. generate tasks based on predefined lists of different elements and thus already receive a certain number of distinct task structures [5]. Cristea et al. developed an algorithm which creates electrical circuits according to user defined properties and generates associated numeric values [6].

An advanced tool for task generation for the theory of mechanics and statics is being developed at the MacEwan University in Canada. Exercises are created according to the type and difficulty level in form of a PDF and additionally generated multi-page solutions with images and algebra, see Fig. 1. For defining the parameters of the tasks C++ is used, for creating the graphics Asymptote and for the set Latex is used [7]. Therefore, it is difficult to implement a web application based on the technologies used. However, the importance and success of such tools is supported by the awards received by the project [8].

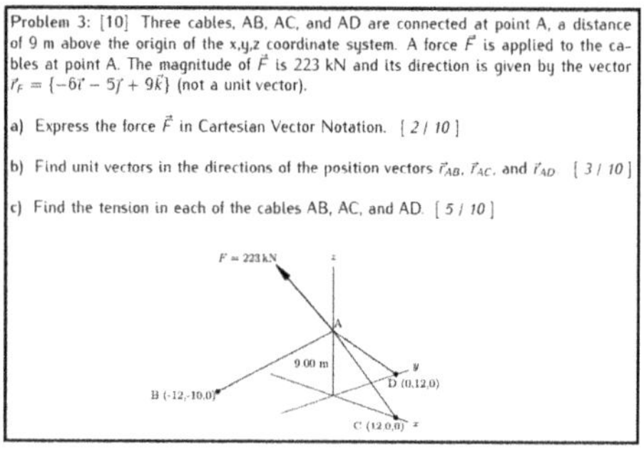
Problem 3: [10] Three cables, AB, AC, and AD are connected at point A, a distance of 9 m above the origin of the x,y,z coordinate system. A force $\vec{F}$ is applied to the cables at point A. The magnitude of $\vec{F}$ is 223 kN and its direction is given by the vector $\vec{r}_F = \{-6\vec{i} - 5\vec{j} + 9\vec{k}\}$ (not a unit vector).

a) Express the force $\vec{F}$ in Cartesian Vector Notation. [2 / 10]

b) Find unit vectors in the directions of the position vectors $\vec{r}_{AB}$, $\vec{r}_{AC}$, and $\vec{r}_{AD}$ [3 / 10]

c) Find the tension in each of the cables AB, AC, and AD. [5 / 10]

Fig. 1. Problem generated by [7]

4 Concept of Tool²Task

Based on the situation and motivation described in Sect. 2 the aim of Tool²Task can be described as follows. The goal is to provide a web based application which automatically generates tasks and makes them available to the students individually. At the same time the tool should generate highly varying and as random as possible exercises.

The non-representative survey of students and the creation of possible user stories have led to the following requirements. Starting from a selection of the type of exercise and the level of difficulty, a random task has to be generated and shown in the web browser, see Fig. 2. At the same time an option to export the task as PDF has to be offered. This enables students to print the task and solve the graphical subtasks. After the first attempt of solving the problem, students want to compare their solution with a short solution, providing the numeric solutions, to check their solution without looking into the complete solution. If this is not sufficient to get the correct solution, only the view in the complete solution helps. However, the complete solution should be shown step-by-step, so that unintended spoiling is avoided, see Fig. 3. For each task a unique ID should be generated, which enables students and lecturers to recreate a task which was generated by another student.

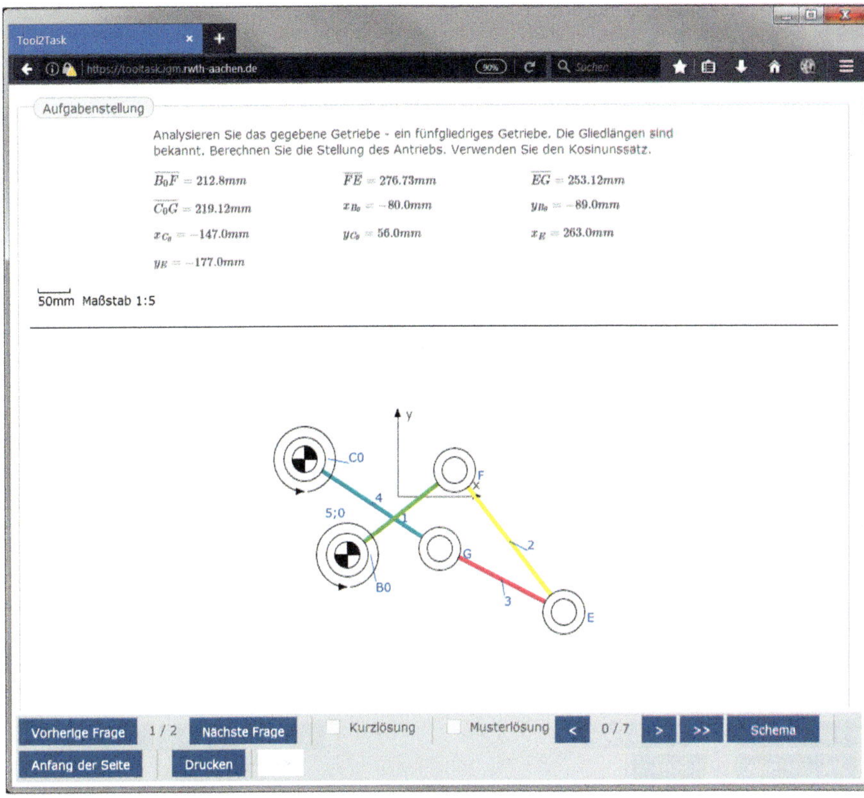

Fig. 2. Generated exercise of pose analysis type

At RWTH Aachen University a unified learning and teaching platform, called L^{2}P, is used to provide learning materials, media and additional information at one place. Therefore, an additional requirement for Tool^{2}Task is to be compatible to this platform and support the given authentication system. This limits the technical implementation to certain web technologies.

5 Created Algorithms for Task Generation

Due to the limited space, the designed algorithms are not presented here in detail. However, an overview and the rough structure are shown, so that an insight into the functioning is given. Furthermore, the approach can be considered the basis for comparable problems.

For the generation of problems, in principle, initially the complete system is calculated. For instance, a kinematic structure is chosen, link lengths and a drive angle generated. Thereafter, a sought variable and several given variables are chosen randomly. In a final step, the task is solved in order to create the complete solution.

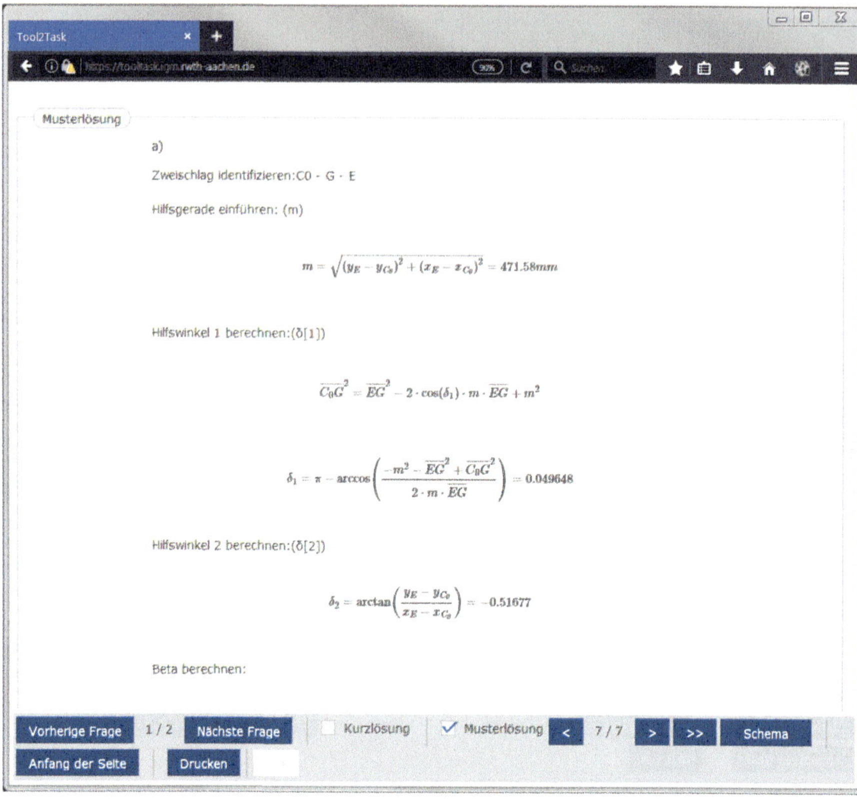

Fig. 3. Assigned solution of task in Fig. 2

Besides mechanism theory also electric drives are treated in EMAT. The calculation of drive trains can be considered as a classical algebraic equation system problem. For this type of tasks, an algorithm was created to generate linear and polynomial equation system problems. Therefore, a complete list of possible equations belonging to the problem category is given. A loop for increasing the equations system is passed through until the desired degree of difficulty is reached.

The first step of the loop is the random selection of an additional equation. Based on all selected equations a list of possible unknowns is created. In the next step, a temporarily sought variable is chosen temporarily. Note, whenever a new equation is selected, also the sought variable might change. Hence, only in the last step of the loop the final unknown is determined. Before the next iteration of the loop a list of possible additional equations is created. For this step various conditions are checked. Is the system solvable with the new equation? Is the new equation linked to the previous equations, or can the equations be solved independently? Note, equation system problems are created straight forward and not by trial and error. Especially the algorithm for generating equation system problems generated genuinely random tasks, of which even the authors have not thought so far.

6 Implementation of Tool^{2}Task

The focus of Tool^{2}Task lies in natural sciences, engineering sciences and especially mechanism theory, where tasks are usually a combination of graphical and algebraic methods. Within these boundary conditions, the data model of the software is modular, so that an adaption for new types of exercises can easily take place. For the implementation of Tool^{2}Task *Eclipse RAP* is used. For this reason, it is possible to program new tasks in Java and make use of many existing extensive libraries. The translation of the Java application into HTML and JavaScript to display a website is automatically done by RAP, compare Fig. 4. By using these widely used technologies, integration into virtually any online learning platform, as used by many universities, is possible. Additional technologies used are *MathJax* to display equations and *Maxima* for the algebraic calculations. The basic services of the software comply with classes for the presentation and creation of tasks. In order to implement a new task type, therefore, lecturers have to program the creation and solving of the task type in the form of an algorithm. Furthermore, required equations and application examples have to be inserted into a database.

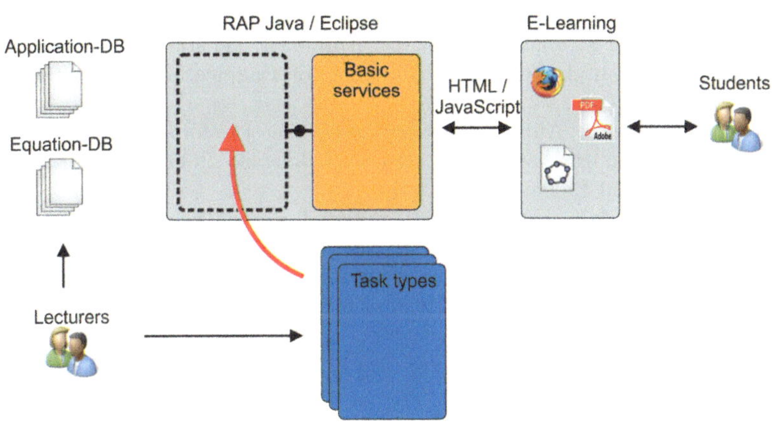

Fig. 4. Implementation of Tool^{2}Task

7 Conclusion and Outlook

With web-based tools for automatic task creation like Tool^{2}Task, it is possible to provide any number of individual exercises. This allows students to better shape their learning process to their needs. In particular, students who need more exercises to practice or already failed one exam are supported.

In the future Tool^{2}Task will be expanded further. The provision of solutions may be linked to the input of a solution by the student. Created tasks could be provided as files for geometry applications like GeoGebra. Furthermore, it would be interesting to provide animated mechanisms at the beginning of the learning process in order to increase the understanding of movements and kinematics. The possibility to show additionally theory instead of solutions might help students to solve the problems themselves.

Acknowledgements. This project was funded by the Exploratory Teaching Space of RWTH Aachen University and the authors would like to thank for the support. Furthermore, the authors thank Fabian Weber and Marvin Ahrens for their contribution.

References

1. Kurtenbach, S., Mannheim, T., Hüsing, M., Corves, B.: Content and realization of education in mechanism theory at RWTH Aachen University. In: New Trends in Educational Activity in the Field of Mechanism and Machine Theory (2014)
2. Körndle, H., Narciss, S., Proske, A., Carstensen, D, Barrios, B.: Konstruktion interaktiver Lernaufgaben für die universitäre Lehre. In: Campus. Kommen die digitalen Medien an den Hochschulen in die Jahre? Waxmann, Münster (2004). ISBN 3-8309-1417-2
3. Fikar M., et al.: Automatic generation of assignments and quizzes in control engineering education. In: Proceedings of the European Control Conference (2007)
4. Kutasoftware (2015). https://www.kutasoftware.com
5. Prados, F., Boada, I., Soler, J., Pocj, J.: Automatic generation and correction of technical exercises. In: International Conference on Engineering and Computer Education (2005)
6. Cristea, P., Tuduce, R.: Automatic generation of exercises for self-testing in adaptive E-learning systems: exercises on AC circuits. University Politechnica of Bucharest, Faculty of Electrical Engineering (2005)
7. Fyfe, K., Davis, J.: Automated generation of randomizable problem sets and detailed solutions in engineering statics. In: Proceedings from NW Regional ASEE Conference (2013)
8. PNW – ASEE Section Meeting (2013). http://pnw.asee.org/Newsletters/2012/ASEE_PNW_2012.htm

Strengthening the Interaction of Art and Science Through Rubric-Based Evaluation Models: The Final Degree Project of the Dual Degree of Engineering in Industrial Design and Mechanical Engineering

R. Díaz-Obregón(✉), S. Nuere, R. D'Amato, and M. Islán

Polytechnic University of Madrid, Madrid, Spain
raul.diazobregon@upm.es

Abstract. Massachusetts Institute of Technology (MIT), one of the most prestigious institutions in technological innovation, incorporates art and creativity as a distinctive element of its research, technological degrees, humanities and creative subjects. Unfortunately, art and creativity do not have the same consideration. This study intends to present a work line corresponding to the art-science integration, which is part of a renovation-didactic-project developed in the study-area of industrial design at the Higher Technical School of Engineering and Industrial Design (EscuelaTécnica Superior de Ingeniería y Diseño Industrial, ETSIDI of the Polytechnic University of Madrid, Spain). The benefits of rubric use in order to facilitate a creative-aspects-evaluation guide in the double degree of Engineering in Industrial Design and in Mechanical Engineering will be analysed precisely.

Keywords: Art and science · Rubric-based evaluation

1 Introduction

The aim of this study is to offer a more up-to-date, fair and quality education to the students, and to prove the benefits offered by the integration of artistic activities in many areas of daily life, and particularly in the scientific-technical environment. Nevertheless, it is important to face the fact that this area of knowledge does not know and does not apply the possibilities offered by creativity and art. The aim of this study is to contribute to the approach of the two environments and to offer objective tools for the evaluation in artistic and creative contents.

In the first step, the working contexts will be defined and some connections between art and science will be explained. Afterwards, the part of the educational line in the design area will be explicated, through the creation of the evaluation rubrics. This study will be outlining what represents an evaluation rubric and the interesting educational possibilities that it offers. The next step will be to analyze the work performed at the High Technical School of Engineering and Industrial Design (ETSIDI) of the Polytechnic University of Madrid and the current interest in the creation of a rubric for

J. C. García-Prada and C. Castejón (Eds.): *New Trends in Educational Activity in the Field of Mechanism and Machine Theory*, MMS 64, pp. 141–148, 2019.
https://doi.org/10.1007/978-3-030-00108-7_16

the Final Degree Project (Trabajo Fin de Grado, TFG) for the double degree in Engineering in Industrial Design and in Mechanical Engineering that contemplates a greater definition of the creative and artistic aspects.

2 Artistic Backgrounds in an Engineering Environment at the ETSIDI

The need is to place the studies of Engineers in Industrial Design in a High Technical School with a strong tradition of engineering and where there is a lack of interest for more humanistic subjects. So the first step in these studies was to introduce subjects with a strongly creative nature and of a clearly differentiated methodological conception, switching from the classic theoretical classes to a concept more related to the creation of projects. As the artistic concepts may be undervalued, it is important an evaluation method that gives credibility to this subject. It can be establishing several antecedents to the teachers' methodological commitment that integrates the interdisciplinary development, the horizontal development, the collaborative work or the connection with the professional world. The teachers have been to introduce new methodologies, mixing this vision of science and art, working together in order to improve the formation training of the students. Art is considering as the development of intellectual activities such as creating or interpreting, encouraging self-expression and perceptive student qualities [1].

The quotation of Ortega y Gasset: "So engineers see how to be an engineer is not enough to be an engineer", introduces the necessity of looking for the approximation of being capable of affording any kind of problem with a wider vision and an added value that could enrich their profession. Engineers will use the technology to make up an idea, a life experimentation so as to generate progress in the society.

At this point it is necessary to establish relationships between art, design and engineering to understand that all kind of knowledge, instrumental (material) or only intellectual (abstract or artistic) will conform what it is known as purely culture. One of the clearest examples where art, design and engineering have advanced hand in hand dates back to the Renaissance, at the end of the 15th century, with the work of Leonardo da Vinci. The Bauhaus is another great movement in Design that has demonstrated the power of working and mixing different fields of knowledge for the benefit of the society. But it is also possible to find the interest of the Massachusetts Institute of Technology to introduce art into its current STEM movement (Science, Technology, Engineering and Math) becoming nowadays STEAM (A standing for Art). Some graduated from this Institute have already told how useful was studying many other subject-matters as history, literature or philosophy to improve the development of their empathy and critical thinking skills [2]. Through the exploration of the humanities it is possible to learn to think creatively and critically, and that it serves above all to ask questions (the basis of all development and improvement). As Helen Longino says [3], the sciences have become hyper-specialized. Scientists rarely have the opportunity or support to step back from their research and ask how it connects with other work on similar topics.

3 Approach to Rubrics Through Art

According to Alsina [4], the rubric is a tool that comes from the fields of psychology and education "where a qualitative object (e.g. text) is related to quantitative objects (e.g., metric units)." He determines that its goal is to offer and share with students a series of evaluation criteria, quantifying and describing the degree of resolution.

Alcón and Menéndez [5] emphasize that this document generally provides "evidence to issue a quality judgment on a student's performance" and contributes to "achieving greater interdependence between teaching, learning and evaluating processes", by generating a more solid didactic proposal. The use of the evaluation rubrics provides an important step towards establishing fairer evaluation criteria and greater communication with teachers, by leading to an improvement in academic results, as Menéndez and Gregori-Giralt point out.

In the authors' opinion, an examination of validity should demonstrate that students have successfully overcome cognitive challenges, and that the learning process has been confirmed by various observers – including students. The highly specialized and contextualized character of the learning outcomes expected in higher education is beyond any doubt. This one is no reason to deny the possibility of generalization of a well-designed assessment system, if this concept is redefined [6].

These materials help to have a complete educational experience, especially in the higher educational levels, in which, the lack of coherent didactic approaches, didactic processes updating and educational university training, together with the one in which the widespread growth, generate very precarious and poorly agreed evaluation systems.

Alsina explains that the rubric "is a resource that expresses in writing the result of the dialogue and consensus of a group of professionals committed to improving the quality of their professional practice" [4]. He adds that, these tools facilitate both the student and the rest of society understanding more complex aspects, by reinforcing their autonomous learning. It is a device that provides concrete data on the marks and on the evaluation of the work done, by orienting the student to the acquisition of new competences. In addition, it is an excellent didactic tool to discuss the score and to bring the didactic processes closer to the student.

The rubrics implementation in the design area at the High Technical School of Engineering and Industrial Design (ETSIDI) is summarized in the use of 4 exercises of the Graphic Design subject, which obtained a great acceptance of the students. However, for years, specific evaluation guides for each exercise have been used, which specified the evaluation criteria for each work. These guides were integrated in a document called exercise guidelines that served as an extension of the teaching guide. However, it has been detected that there is an alarming dearth of specific materials to create and evaluate the Final Degree Project, which has forced us to prioritize this field.

4 Final Degree Project (TFG)

The final-grade work, project or subject, as it appears in the Spanish Official State Gazette must "be carried out in the final phase of the curriculum and be oriented to the evolution of competences associated with the degree". However, the legislation also leaves a wide margin of action for each university to generate its own definition [7].

Due to the diversity of study programs in which the number of teachers involved and the specificity of each work are coordinate [8], it is especially difficult to define specific materials. This situation leads, in many cases, to an almost total absence of materials, to deficiencies in the direction of the works and also to the irregularity of the evaluation criteria.

In this sense Stapé-Dubreuil recommends the creation of generic master sessions on documentation, organization, and presentation of the TFG. He also determines the usefulness of regulating and structuring the tutoring sessions, by specifying the minimum dedication in each part of the use of the teaching guides, in which they are collected. But what is necessary in the TFG, as in any didactic process, is the definition of an evaluation guideline.

In ETSIDI School, there is no information that facilitates the work to direct, to structure, to organize, to develop, to present or to evaluate one of the most important subjects of the degree. The instructions are reduced to seven points that refer to norms of formal cover-presentation, of format and of electronic display, but do not explicitly integrate the mentioned elements. This circumstance leads to a very heterogeneous criteria application and to a highly irregular development of teaching and evaluation processes that, from a didactic point of view, are not correct.

5 The Rubric in the TFG

As noted above, there are materials and actions that facilitate the elaboration of the TFG. Firstly, the most common is the creation of detailed rules which explain the whole process of enrolment, selection of the subject, tutors' assignment and evaluation processes. The second step is the writing of evaluation guides, style manuals or index suggestions. In some cases, and not less recommendable, as Stapé-Dubreuil [8] points out, is the teaching of lectures and the preparation of evaluation guides. Less common are the TFG monitoring and mentoring sheets.

However, the most controversial aspect remains the evaluation and marking (qualification) process. In this aspect, the most interesting material as it has pointed out above and as Stapé-Dubreuil writes down, are the evaluation rubrics. It is recommended "the use of specific rubrics designed to ensure, as far as possible, an objective final marking (qualification) process." In addition, it is emphasized that the rubric must contemplate the formative processes that are to evaluate the processes and not only the results. Lastly, it must be configured to "allow the revision of a given grade if a student requests it" [8].

6 A Rubric Development with Artistic Components

The legal framework of the TFG determines the need to evaluate the competences of the degree collected in the National Quality Assessment and Accreditation Agency of Spain (ANECA). In the double degree of Industrial Design Engineering and in Mechanical Engineering, most of the defined criteria facilitate their compilation in a rubric, but it generates an insurmountable ambiguity in certain aspects (Fig. 1).

General Competence Number	Competence
CG 1	To know and to apply a basic science and technology
CG 2	To have the capacity to design, to develop, to implement, to manage and to improve products, systems and processes, by using appropriate analytical, computational or experimental techniques.
CG 3	To apply the acquired knowledge in order to identify, to formulate and to solve problems in broad contexts, by being able to integrate them working in multidisciplinary teams.
CG 4	To understand the impact of engineering on the environment, on the sustainable development of society and on the importance of working in a professional and responsible environment.
CG 5	To communicate knowledge and conclusions, in oral, written and graphic way, to specialized and unspecialized audiences in a clear and unambiguous manner.
CG 6	To have the learning abilities that allow to continue studying throughout the life for a suitable professional development.
CG 7	To incorporate the ICT and the technologies and tools of the Engineering in Industrial Design and Product Development in their professional activities.
CG 8	To develop the ability to work in a bilingual environment (English - Spanish)
CG 9	To organize and to plan projects, and human teams. Teamwork and leadership skills.
CG 10	Creativity

Fig. 1. General competences

Attention must be paid to the fact that the degree of competence number 10 (CG10), common to both grades, obtains the descriptor "creativity" without giving any explanation. Although this competence is tremendously confusing in the field of creativity and art. It generates a lack of definition for non-specialized environments. For this reason, and for those exposed in previous points we considered the definition of these sections opportune.

In this second study phase it has been proceeded to compile materials, both specific to the study of Industrial Design Engineering and Mechanical Engineering, as well as to degrees with more creative teachings, especially Fine Arts. In this process is possible to highlight two aspects:

- There are very few published materials on specific rubrics in these two areas.
- There is no greater scope in the scientific areas than in the artistic areas.
- Despite the apparent differences in study areas, the proposed structures are very similar.

As mentioned above, it is possible to see that in the materials with a certain quality, some TFG development outlines were offered and these were very similar, always appreciating an introduction, a development, conclusions, formal aspects and oral presentation guidelines. In their comparison it was possible to analyse the strengths in different aspects, being very feasible their combination and adaptation to our particular case.

7 A Rubric Development with Artistic Components

Based on the previous studies and by taking into account the degree competences approved by ANECA, it was possible to establish the following scheme, in which it is appreciable both the structure of the work of the TFG as well as the sections of the rubric. In the same case it was possible to appreciate their correspondence with the degree competences and, lastly, the value of each section (Fig. 2).

1. MEMORY		**70%**
1.1. Index, introduction and study definition	CG 8	15%
1.2. Development		30%
1.2.1. Concept, process and creativity	CG 2, CG 10	15%
1.2.2. Technical and scientific aspects	CG 1, CG 2, CG 7	15%
1.3. Results and social, environmental and commertial impact	CG 3, CG 4, CG 9	15%
1.4. Conclusions and continuity	CG 5	10%
2. FORMAL ASPECTS AND COMMUNICATION		**30%**
2.1. Academic writing	CG 5	10%
2.2. Design and images	CG 5	5%
2.3. Presentation and oral defense (part in English)	CG 5, CG 8	15%

Fig. 2. Sections of the rubric, degree competences and grading percentages

Due to the great extension of rubric and by following a thematic coherence, in the section "1.2.1 Concept, process and creativity", it was possible to analyse the competence of degree number 10 Creativity, as reported in the above table.

Within this section the creative concept has been incorporated, which alludes to the intellectual capacity to defend a creation, by arguing its most significant referents. In both, the artistic and the scientific fields, a product is not conceived without prior analysis, and a new contribution is presupposed to that which already exists.

Another very important aspect, usually forgotten in educational scopes and of essential importance in artistic-creative teaching, is the recording of creative processes. It is considered a necessity that the student of design has acquired tools to visually and verbally express the stages of its creation, by giving coherent and critical explanations of all its phases of study.

Finally, the section creativity relates to the concept of originality and innovation and is associated with the previous two. The student is expected, firstly, to thoroughly review available sources, then to develop a broad study of possibilities, and finally, to provide a different, intelligent, meaningful and innovative solution for their educational level.

Although the creative processes are much more complex and broad than those presented here, it is consider that these three elements summarize the minimum competences that should be required of students of Engineering in Industrial Design and in Mechanical Engineering.

8 Conclusions

The initial objective of this study was to show the potential offered by the integration of artistic activities in the scientific and technical environment, by indicating the confluences of the two environments and by exposing a practical case of rubric use developed at the ETSIDI.

The Final Degree Project is a mandatory study aimed at consolidating the competencies of each grade, whose development is very irregular. Stapé-Dubreuil [8], points out the difficulty of implementing the TFG and recommends a series of actions for its achievement among which was the use of rubrics. Despite the importance of this mandatory academic work, the degree of the ETSIDI does not have the specific materials for the management, development and evaluation of the TFG. On the other hand, and unlike other competences of the degree, there is no guideline that defines the creative competition as a situation that requires an urgent action.

The degree-rubric-configuration is based on other sections of TFG in both scientific and technological fields. In these areas there was a significant lack of materials and a very important similarity of structure and contents. These confluences facilitated the creation of a specific structure. On the other hand, an adaptation of the degree competences to this scheme has been performed, by facilitating a schematic proposal, in which both the competences of degree and the proportions suggested for the qualification appear.

Finally, it was possible to define the competition number 10 creativity, by using the references of the Fine Arts and Engineering rubrics studied. In this respect and based on the confluence of both areas three aspects have been defining: the conceptual definition of creation based on previous referents, the registration of creation processes and, finally, originality and innovation, which are related to documentation and conceptual definition.

Once the process is completed, the created materials can contribute to the improvement of educational quality, by facilitating materials with less didactic fissures and fairer. Likewise, the offered data can help to establish bridges of union with creative-artistic matters, as well as to provide more concrete guidelines for evaluating creativity in technical-scientific studies. Finally, it is possible to affirm the quotation of Robin Mathew that well defines purposes of this study: "Design is where science and art reach a point of equilibrium".

References

1. Nuere, S.: El arte y el diseño en la ingeniería. Rev. UPM **32**, 29–32 (2015)
2. Herranz, A.: Menos STEM y más Platón. En el futuro de la tecnología, los estudiantes de humanidades serán muy buscados. Vida Tecnológica (2016). http://www.bloglenovo.es/humanidades-importantes-tecnologia-triunfar-futuro. Accessed 12 Jun 2017
3. Longino, H., Wilcox, B.: Simple isn't better when talking about science, Stanford philosopher suggests. The Humanities at Stanford (2014). http://news.stanford.edu/news/2014/july/aggression-philosopher-longino-072514.html. Accessed 12 Jun 2017
4. Alsina, J.: ¿Qué es una rúbrica?¿Por qué y cómo se utiliza? In: Alsina, J. (ed.) Rúbricas para la evaluación de competencias, pp. 8–13. Cuadernos de docencia universitaria. Octaedro, Barcelona (2013)
5. Alcón, M., Menéndez, J.L.: La contribución de las rúbricas a la práctica de la evaluación auténtica. Observar **9**, 5–17 (2015)
6. Menéndez, J.L., Gregori, E.: The contributions of rubrics to the validity of performance assessment: a study of the conservation-restoration and design undergraduate degrees. Assess. Eval. High. Educ. **41**(2), 228–244 (2016). http://www.tandfonline.com/doi/abs/10.1080/02602938.2014.998169. Accessed 18 May 2017
7. Ferrer, V., García-Borés, J.: El TFG: miradas y retos. In: Ferrer, V., Carmona, M., Soria, V. (eds.) El trabajo de fin de grado: guía para estudiantes, docentes y agentes colaboradores, pp. 1–39. Mc Graw Hill-Interamericana de España, Madrid (2013)
8. Stapé-Dubreuil G et al.: Trabajos fin de grado: la coordinación como clave del éxito. Available via CIDUI (2012). http://www.cidui.org/revistacidui/index.php/cidui/index. Accessed 15 June 2017

Analytical or Computer-Aided Graphical Methods for Introductory Teaching of Mechanism Kinematics?

Pietro Fanghella(✉), Giovanni Berselli, and Luca Bruzzone

DIME, University of Genoa, Genoa, Italy
{pietro.fanghella, giovanni.berselli, luca.bruzzone}@unige.it

Abstract. The paper presents, through an example, two different approaches to introduce mechanism kinematics to first-cycle mechanical engineering students. Analytical solution and traditional graphic methods are considered, both supported by state-of-the-art software tools. Advantages and disadvantages of both approaches from an educational point of view are outlined during example development. Finally a summarizing discussion is proposed.

1 Introduction

In Italian universities, almost all first-cycle degrees in mechanical engineering include, usually in their second year, a first course on mechanics of machines within which an introduction of kinematics and statics of mechanisms is presented [1]. Typically, the expected practical learning outcomes are the capability to analyze the kinematic behavior (position, velocity, acceleration) of closed loop mechanisms, the operative understanding of the concepts of mobility and degrees of freedom and the consequent choice of independent variables, and the acquisition of introductory notions regarding configurations and singularities. Starting from the properties of rigid bodies and kinematic pairs, the teaching track usually presents methods for the formulation of kinematic models and for their solution. Then a certain amount of application examples developed in detail, in order to allow students to apprehend both theory and practical issues (see for example [2], chapters 2 and 4 for typical reference material).

Focusing on application examples, that represent a key element of learning, there are many possibilities to implement them, and the choice of the approach drives the software tools used for their development. One possibility is to adopt a mixed algebraic-geometric approach [3], based on a well-known educational tool for teaching math and geometry [4]. Other possibilities are either pure algebraic approaches based on computer algebra tools, or more geometrically oriented solutions, based on CAD systems [5–7]: here, through an example, we will discuss and compare these two approaches, the former supported by Maple software tool [8], the latter by using a constraint based parametric 2D CAD software with embedded constraint solver [9, 10] and available free of charge [11]. The two adopted approaches and tools are widely used to solve engineering problems, so, while learning kinematics, students have also the opportunity to tackle and learn engineering tools within the process of solving

J. C. García-Prada and C. Castejón (Eds.): *New Trends in Educational Activity in the Field of Mechanism and Machine Theory*, MMS 64, pp. 149–156, 2019.
https://doi.org/10.1007/978-3-030-00108-7_17

engineering problems. Problem solving by the use of complex modelling tools is a key competence of a young mechanical engineer, as in many cases this kind of skill is needed in actual working situations.

The paper presents an example problem, then its solution by means of the two approaches is discussed in details, with specific reference to difficulties and learning opportunities. Finally conclusions are drawn.

2 Example Mechanism

A well-known mechanism is chosen for the discussion, the Roberts straight line linkage, shown in Fig. 1: the mechanism is a four bar, whose geometry requires that links 2 and 3 and segments BP and CP on coupler link 4 have the same length, while the length of frame link 1 is twice the length of the segment BC on link 4. To develop the mechanism model, a reference frame is positioned as in Fig. 1a, with its origin O coinciding with coupler point P when the mechanism is in its symmetric position.

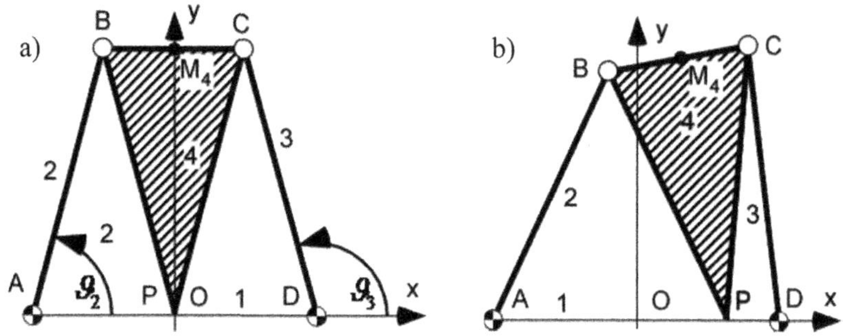

Fig. 1. Roberts straight line linkage

When the mechanism is moved in both directions from its reference symmetric configuration of Fig. 1a, the coupler point P traces a trajectory that for a certain motion range is very close to a straight line coinciding with the x axis (Fig. 1b).

A typical application problem to be presented to students, after the discussion about fundamentals of kinematic analysis, can be:

- to determine the trajectory (x_P, y_P) of point P, and consequently the error (y_P) of the mechanism in tracing a straight line motion
- to determine the angular velocities of the links and the velocity components of point P.

The problem formulation allows students to reason about the difficulties related to closed-loop position analysis, with the determination of the coupler point trajectory and velocity, and, at the same time, to consider the functional aspects of the linkage and how they are related to its dimensional design. In addition, students are led to observe that in kinematics problems approximate solutions are often very useful.

3 Solution of Example Mechanism: Analytical Approach

Traditionally, the problem introduced in the previous section can be solved by setting up and solving a mathematical model of the linkage. In order to outline the advantages and disadvantages of this approach from an educational point of view, the problem solution will be developed in the following.

Assuming $l_2 = l_3 = l_{4BP} = l_{4CP}$ (l_2 hereafter) and $l_{1AD}/2 = l_{4BC}$ (l_4), the linkage closure may be stated by imposing a relation on links 2 and 3 and the length of l_4:

$$(B - C)^2 - l_4^2 = 0 \tag{1}$$

in which $B = \begin{bmatrix} -l_4 + l_2\cos(\vartheta_2) \\ l_2\sin(\vartheta_2) \end{bmatrix}$ and $C = \begin{bmatrix} l_4 + l_2\cos(\vartheta_3) \\ l_2\sin(\vartheta_3) \end{bmatrix}$

By substituting and simplifying (the use of a computer algebra program such as Maple is very helpful in this activity), the following closure equation, relating angles ϑ_2 and ϑ_3, is obtained:

$$\begin{aligned} &\left(4l_4l_2 - 2l_2^2\cos(\vartheta_2)\right)\cos(\vartheta_3) + \left(-2l_2^2\sin(\vartheta_2)\right)\sin(\vartheta_3) \\ &\quad + \left(-4l_4l_2\cos(\vartheta_2) + 3l_4^2 + 2l_2^2\right) = 0 \end{aligned} \tag{2}$$

Equation (2) can be rewritten in the well-known form:

$$U(\vartheta_2)\cos(\vartheta_3) + V(\vartheta_2)\sin(\vartheta_3) + W(\vartheta_2) = 0 \tag{3}$$

and it can be solved by the expression [2, App. A]:

$$\vartheta_3 = 2\arctan\left(\frac{V \pm \sqrt{V^2 + U^2 - W^2}}{(U - W)}\right) \tag{4}$$

Substitution into Eq. 4 of explicit expressions of coefficients U, V and W (see Eq. 2) yields a complicated and error prone expression, so, in order to obtain numerical results and plots, students should be suggested to adopt either a computer algebra tool (e.g., Maple) or to formulate the problem with an algorithmic approach (Matlab, Excel) allowing them to compute numerical results from Eq. (4) through a step by step method.

Once angle ϑ_3 is obtained (for the considered mechanism by adopting the "+" solution of Eq. (4)), angle ϑ_4 can be obtained by the following relation:

$$\vartheta_4 = \arctan 2\left(\left(C_y - B_y\right), \left(C_x - B_x\right)\right) \tag{5}$$

yielding:

$$\vartheta_4 = \arctan 2((l_2\sin(\vartheta_3) - l_2\sin(\vartheta_2)), (2l_4 + l_2\cos(\vartheta_3) - l_2\cos(\vartheta_2))) \tag{6}$$

Finally, the coordinates of point P as a function of angles ϑ_2 and $\vartheta_4(\vartheta_2)$ are determined (point M_4 is the midpoint of segment BC, $l_{4h} = \sqrt{l_2^2 - (l_4/2)^2}$ is the length of segment M_4P, Fig. 1):

$$P = M_4 + (P - M_4) = \frac{(C+B)}{2} + l_{4h}\begin{pmatrix} \sin(\vartheta_4) \\ -\cos(\vartheta_4) \end{pmatrix} \quad (7)$$

Again, the explicit substitution of ϑ_3 solution into Eqs. (5), (6) and (7) is very impractical, so resorting to some form of software tool and implementation is required.

Having assumed $l_2 = 180$ mm and $l_4 = 120$ mm, the plots in Fig. 2, obtained by Maple, show some numerical results of position analysis: (a) the angle $180 - \vartheta_3$ whose behavior is symmetrical to that of ϑ_2, the two angles being equal in the reference configuration of Fig. 1a, (b) the trajectory (x_P, y_P) of coupler point P, in which the y coordinate represents the error with respect to exact straight line tracing; please note that axes in Fig. 2b have very different scales, so a student could appreciate that the tracing is approximate, but with an error lower than 1 mm for a range -130 mm $< x_P <$ 130 mm.

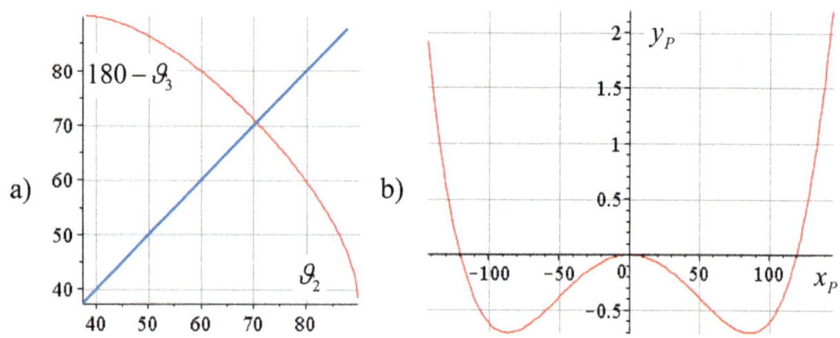

Fig. 2. Plots of $180 - \vartheta_3$ (deg) and (x_P, y_P) (mm)

The velocity equation to determine ω_3 can be easily obtained by deriving Eq. 2, while ω_4 is less easily obtained either by deriving Eq. 6 or by solving the first system in Eq. 8 (see next Section); here, thanks to Maple features, the direct differentiation is adopted. Plots in Fig. 3 show ω_3 and ω_4 as functions of ϑ_2, for $\omega_2 = 1$ rad/s.

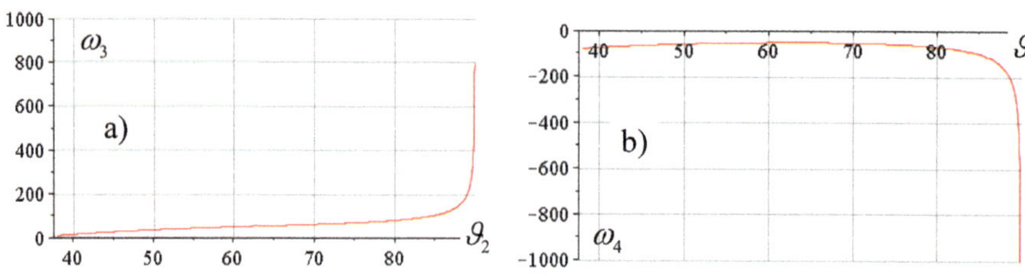

Fig. 3. Plots of ω_3 and ω_4 (deg/s)

4 Solution of Example Mechanism: Computer-Aided Graphical Method

Before computational approaches were widely adopted due to the increasing availability of suitable computational tools, graphics methods were generally used to study mechanism kinematics. The main advantages of such approaches are the relative simplicity and the clear geometrical meaning and insight, both features being very valuable for introductory educational steps. Unfortunately, one key aspect of graphic methods is that “hand-made” graphic constructions or even traditional CAD based approaches provide only static pictures of the mechanism position and velocities, so that many drawings are necessary to get an overall comprehension of the mechanism characteristics. To overcome this drawback, without resorting to complex 3D modelling tools that may not be suited to second-year students, an interesting solution is provided by 2D parametric CAD tools equipped with embedded constraint solvers, as briefly discussed in Introduction. Almost all parametric 3D software products (e.g., PTC Creo Parametric, Dassault Catia, Siemens NX, Dassault Solidworks, Siemens Solidedge) contain a software module, here generically named “sketcher”, devoted to the sketching of 2D sections, shapes, curves and other planar geometric entities useful to build complex 3D geometries according to a constructive approach. All parametric sketchers share some interesting characteristics, also useful in graphical methods for planar mechanism kinematics and statics:

- it is possible to apply geometric constraints to the sketched objects; such constraints (e.g., horizontal, vertical, coincident, tangent, orthogonal, parallel,…) are permanently enforced and constrain the sketch variations; in other terms, they build up the constraint manifold of the mechanism;
- similarly, it is possible to add dimensional constraints, i.e., “driving dimensions” (e.g., lengths, angles, radii, equality between dimensions,…) whose values determine the sketch size; moreover, the sketcher allows to add “driven dimensions”, i.e. dimensions that do not constrain the sketch, but are useful to measure quantities as a function of the driving dimensions;
- the sketcher computes the number of degrees of freedom of the sketch on the basis of imposed geometric and dimensional constraints, indicating when the sketch is fully determined (zero d.o.f.);
- the sketcher recognizes conflicts among constraints and (usually) asks the user to choose a set of non-conflicting, independent constraints.

Although sketcher software modules are nowadays a well-established component of several 3D modelling tools, it is less easy to find stand-alone 2D software tools with the previous features: here we use the product “Solid Edge 2D Drafting” from Siemens [11], which is a 2D subset of the 3D Solid Edge. For our educational purposes it is relevant the fact that Solid Edge 2D is freely distributed by Siemens. Moreover Solid Edge 2D, beyond fully supporting parametric and constraint based functions, is also a featured 2D engineering drawing tool.

In order to create the model of the chosen mechanism within Solid Edge 2D, a rough sketch is first created, then dimensional and geometric constraints are added, and finally driven dimensions (e.g., those required to measure the position x, y of point P). Figure 4a shows the Roberts mechanism in its reference position with three driving dimensions (l_1, l_2 and l_{4BC}, drawn in black), equality constraints imposed so that $l_2 = l_3 = l_{4BP} = l_{4CP}$ and various driven dimensions (drawn in light blue in figure), in particular the angles ϑ_2 and $180 - \vartheta_3$, and the two coordinates of P. According to its constraints, the sketch in Fig. 4a has one d.o.f., so it can be moved to ∞^1 different positions, for example by dragging the vertical green segment, connected to point P, that was purposely created as a handle to move the mechanism. Figure 4-b shows the mechanism in the position with $x_P = 50$ mm; the configuration has been obtained by transforming the driven dimension x_P (Fig. 4a) into a driving one and assigning to it the value 50. This change eliminates the remaining d.o.f., so the sketch in Fig. 4b is geometrically determined. Such possibility of switching back and forth the status of dimensions between driving and driven is very useful for educational purposes, as it allows student to become acquainted to the concepts of degrees of freedom and free coordinates; moreover, it allows to freely change the independent variables of the mechanism. This kind of flexibility is much more complex to be obtained with analytical approaches, as a change of the independent variables requires to determine a new solution of the mathematical model.

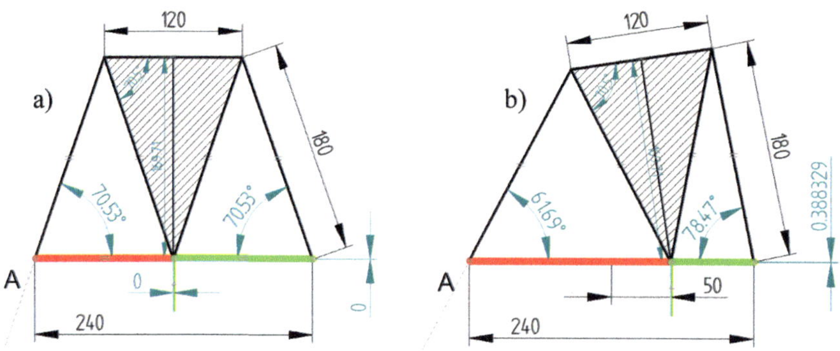

Fig. 4. Parametric sketch of Robert linkage (Color figure online)

A second sketch (Fig. 5, the red circles indicate vector arrows, velocities are expressed in mm/s), related to the position one, can be drawn to solve the velocity analysis problem according to the following relations:

$$
\begin{cases}
\underline{v}_B = \underline{\omega}_2 \times (B - A) \\
\underline{v}_C = \underline{v}_B + \underline{\omega}_4 \times (C - B) = \underline{v}_B + \underline{v}_{CB} \\
\underline{v}_C = \underline{\omega}_3 \times (C - D)
\end{cases}
\qquad (8)
$$

$$
\begin{cases}
\underline{v}_P = \underline{v}_B + \underline{\omega}_4 \times (P - B) = \underline{v}_B + \underline{v}_{PB} \\
\underline{v}_P = \underline{v}_C + \underline{\omega}_4 \times (P - C) = \underline{v}_C + \underline{v}_{PC}
\end{cases}
$$

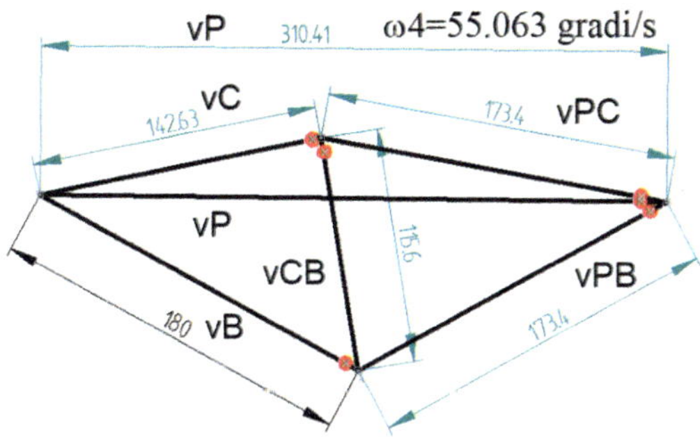

Fig. 5. Parametric sketch for graphic solution of velocity analysis (mm/s) (Color figure online)

The drawing in Fig. 5 corresponds to the position in Fig. 4b, with velocities computed for an input velocity $\omega_2 = 1$ rad/s. Geometric constraints between position and velocity sketches assure that the velocity sketch follows the mechanism position; for example, in Fig. 5, the imposed constraints are: $\underline{v}_B$ orthogonal to link 2, $\underline{v}_{CB}$ orthogonal to segment CB, $\underline{v}_C$ orthogonal to link 3, and so on. For 1 d.o.f. mechanisms, if the correct set of constraints is imposed, the velocity sketch becomes dependent on only one driving dimension, the velocity $\underline{v}_B = 180\ mm/s$ in our case. On the basis of velocity polygon in Fig. 5, also the modules of angular velocities can be obtained by explicit relations added to the sketch: here, for example, ω_4 is obtained as v_{PC}/l_{4PC} and then displayed in the sketch with a variable embedded in a field text. Unfortunately, since dimensions in sketches are in most cases computed as positive numbers, a more complicated approach is required to determine angular velocity signs.

5 Discussion

The two considered approaches to solve the mechanism kinematics have different features that must be evaluated bearing in mind their learning goals:

- *complexity:* the analytical formulation of the problem (Eqs. 1 and 2) is rather straightforward, but its solution, even if it does not require sophisticated mathematical skills (just the knowledge of Eqs. 3 and 4), is moderately cumbersome and there is the risk that students focus on low value algebraic manipulation activities rather than on understanding the key aspects of the mechanical problem; on the other hand, once the basic elements of the selected 2D CAD tool have been learned, the position sketch is usually very simple; a clear comprehension of the velocity problem is required in both approaches;
- *availability of results:* each of the two approaches is suited to best provide specific results; for example, the plots in Figs. 2 and 3 are rather easily created within Maple, provided that the model is solved, while they would require some complicated effort in Solid Edge; on the other hand, interactive graphic restitution of the mechanism motion and velocity vectors, easily obtained in Solid Edge, would be difficultly obtainable in Maple. Another relevant aspect, already mentioned, is that

reversing mechanism input and output variables is simple in Solid Edge, while very costly with the other approach.

In conclusion, from our experience, the graphic approach is best suited to teach introductory mechanism kinematics, but the development of the analytical point of view and the reciprocal validation of the two models may greatly enrich the student experience.

References

1. Bonsignorio, F., Bruzzone, L., Fanghella, P.: Toward a curriculum in mechatronics: two experiences in Italy and Spain. In: García-Prada, J., Castejón, C. (eds.) New Trends in Educational Activity in the Field of MMT, Mechanism and Machine Science. Springer, Berlin (2014)
2. Callegari, M., Fanghella, P., Pellicano, F.: Meccanica Applicata alle Macchine, Città Studi Edizioni (2016). ISBN: 9788825174113 (in Italian)
3. Iriarte, X., Aginaga, J., Ros, J.: Teaching mechanism and machine theory with GeoGebra. In: García-Prada, J., Castejón, C. (eds.) New Trends in Educational Activity in the Field of MMT, Mechanism and Machine Science. Springer, Berlin (2014)
4. Geogebra geometry (2017). https://www.geogebra.org/geometry. Accessed 06 Sept 2017
5. Logue, L.J.: Teaching mechanism design using constraint based design tools in a cad system. In: ASEE 2000 Annual Conference (2000). ISSN: 2153–5965
6. Kurtenbach, S., Prause, I., Weigel, C., Corves, B.: Comparison of geometry software for the analysis in mechanism theory. In: García-Prada, J., Castejón, C. (eds.) New Trends in Educational Activity in the Field of MMT, Mechansim and Machine Science. Springer, Berlin (2014)
7. Corves, B.: Computer-aided lectures and exercises: Graphical analysis and synthesis in mechanism theory. In: Proceeding of the 2004 the Eleventh World Congress in Mechanism and Machine Science. China Machine Press, Beijing (2004)
8. Maple (2017). https://www.maplesoft.com/products/Maple/. Accessed 06 Sept 2017
9. Bettig, B., Hoffmann, C.M.: geometric constraint solving in parametric computer-aided design. J. Comput. Inf. Sci. Eng. **11**(2), 021001 (2011)
10. D Dimensional Constraint manager: Siemens PLM Software (2017). https://www.plm.automation.siemens.com/en/products/open/d-cubed/2ddcm/index.shtml. Accessed 06 Sept 2017
11. Free Solid Edge 2D Siemens PLM Software (2017). https://www.plm.automation.siemens.com/en/products/solid-edge/free2d/. Accessed 06 Sept 2017

Automatic Grading of Student-Specific Exercises in Large Groups of the Subject Theory of Machines and Mechanisms

R. Chamorro(✉), D. García-Vallejo, J. Martínez-Reina, and E. Reina-Romo

Departamento de Ingeniería Mecánica y Fabricación,
Escuela Técnica Superior de Ingeniería, Universidad de Sevilla,
C/. Camino de los Descubrimientos S/N, 41092 Sevilla, Spain
{rchamorro,dgvallejo,jmreina,erreina}@us.es

Abstract. This study establishes an already defined and tested method to grade exercises of kinematics and dynamics within the course entitled "Theory of Machines and Mechanisms" of the Industrial Technologies Engineering Degree at the Seville School of Engineering (Spain). Particular emphasis is made on the automation of grading and personalization of the exercises, due to the large number of students enrolled in this course. The former is made through a teaching platform available at the University of Seville and called Doctus, whilst the latter is achieved by defining the input data of the exercises and the requested results as a function of the digits of the student's ID. The students must face and solve a personalized problem by their own with the knowledge and competences acquired during the academic course. This paper describes the exercises and the tools used to grade them and shows the satisfactory results obtained with these exercises after three academic courses.

1 Introduction

The course on "Theory of Machines and Mechanisms" is basic in the formation of a Mechanical Engineer. In the Seville School of Engineering this subject is taught during the second course in the Industrial Technologies Engineering Degree ("Grado en Ingeniería de las Tecnologías Industriales"). The course covers 6 ECTS and its syllabus was designed to provide the basic knowledge about kinematics and dynamics of planar mechanisms, some concepts of synthesis of mechanisms and some basic aspects of calculation and design of several machine elements (cams, belts and gears). The text book followed has been written by professors of the department [1] and covers all topics treated during the course. Moreover, the syllabus was designed to comply with the Bologna Process for the European Higher Education Area.

Concerning kinematics and dynamics of mechanisms, which spans about 40% of the course, the objective is that students acquire the basic concepts to analyse and design planar mechanisms and understand how the movements and forces

J. C. García-Prada and C. Castejón (Eds.): *New Trends in Educational Activity in the Field of Mechanism and Machine Theory*, MMS 64, pp. 157–164, 2019.
https://doi.org/10.1007/978-3-030-00108-7_18

are transmitted within the mechanism to perform the task it was designed for. The particular objective pursued by the kinematic analysis is understanding the constraints imposed by the joints and visualize the movements of the different links of the mechanism. The objective of the dynamic analysis is to provide the students with the tools needed to associate movements with the forces that cause it or viceversa.

It was detected that, in general, the students have difficulties when facing up to kinematic and dynamic analysis of mechanisms, due to the complex mathematical and physics concepts involved as well as to a likely deficient background. This issue was also observed before the Bologna Process, when this course was offered under a different name, "Machines Theory". After the Bologna process came into effect, the contents of the course were extended and the time devoted to kinematics and dynamics of mechanisms shortened, which increased the difficulties encountered by the students in this part of the subject.

Aware of the problem, the faculty designed an innovative teaching activity to improve the results and, overall, to force the students to learn the contents of the course in a steady and continuous way. The activity proposed a couple of exercises, respectively, a kinematic and a dynamic analysis of the corresponding mechanisms, as a way of self-assessment of their knowledge before the final exam. Due to the fact that the student could solve the exercises at home, these exercises needed to be personalized in order to minimize cheating. It was a real challenge to design personalized exercises for all the students enrolled on this course (around 500) and, more importantly, to grade them quickly. An automated process was clearly required. Another advantage of personalized exercises is that it tries to force the students to solve the problem on their own.

The finally designed activity is presented in this paper, which intends to present an example of personalized problems and automated grading within the Mechanical Engineering curriculum. Automation was achieved with the help of Doctus (http://doctus.us.es/sdocencia/public/index.php), a teaching app developed by the Department of Systems and Automatic Engineering of the University of Seville. Personalization was made by defining the input data of the exercises and the requested results as a function of the digits of the student's ID.

Personalized and automatically graded exercises generated the following benefits:

- The exercises can be more complete and complex.
- They can replace the traditional system of evaluation.
- It is recommended that students use a mathematical software to program the resolution of the exercise, therefore improving their ability in programming engineering like codes.
- They can be used by the student as a self-assessment technique.
- They can help the students to follow the subject up-to-date, guiding and enabling them to learn through practice, which is crucial for students of mechanical engineering.

The next sections are organized as follows: Sect. 2 explains the general methodology used in the design of the exercises to minimize some problems that might occur, such as singular configurations, non-existence of solutions, etc. Section 3 shows two examples: a kinematic and a dynamic exercise, respectively. Section 4 is devoted to results and conclusions obtained from the study of the correlation between the grades obtained in the exercises and those obtained in the final exam.

2 Material and Methods

This section describes the procedure followed to pose personalized, or student-specific, problems to a group of the subject dealt with in this paper. In principle, it is based in two important parts that will be briefly described next. The first one is the web application Doctus that facilitates the management of the personalized problems and the grades generated after the automatic correction of the results submitted by the students. The second important part is the problem itself that has to be well parameterized in order to make it dependent on each student personal data.

2.1 The Teaching Assistive Application Doctus

The web application Doctus was developed in the department "Ingeniería de Sistemas y Automática" of the University of Seville as a platform intended to assist teaching activities in large groups of students. It has been the subject of some papers and a Doctoral Thesis [3]. The application is offered free to any interested university and is continuously maintained by personnel of the department.

The application Doctus is built up with two different interfaces: the teacher interface and the student interface. Both interfaces are accessed by using a web browser and relay on a big data base that stores the information of the students, the exercise the do, the grade they achieve, etc. Among many other utilities, the teacher interface allows the following actions that are useful for the purpose of the evaluation process reported in this paper:

1. To load a list of students pertaining to a group, no matter how large the group is. It is possible to have different groups of students associated to different subjects.
2. To create exercises that can be assigned to a group. Each exercise is composed of a problem statement and a code that will be lately used to correct the results provided by students. In this particular case, the code is programmed using Matlab [2]. This is the key point of Doctus for the purpose of this paper. Since the code can easily account for variables that are dependent on the personal data of a student, the problem statement can be parameterized so that each student will have to solve a different problem.
3. To automatically order the evaluation of the numerical results submitted by the students. For the evaluation MATLAB is called in batch mode to run the same solution code for the data of each stundent. Since the evaluation is done over numerical results, a tolerance of 5 % is typically allowed.

4. To automatically collect the results of all students on a csv file. It is possible to include in the same file the results of all exercises posed to a group.

The student interface need no detailed description. It is enough to say that this interface allows the student to access an exercise and send his numerical results to the data base that is accessed for evaluation. Of course, all students were aware of being evaluated by the application.

2.2 Design of Exercises

The design of the exercise is not a simple task since it should guarantee that each student has to solve a different problem, meaning different known numerical data leading to different numerical results. The rules to design an exercise are summarized here. The exercises referred in this paper focus on different topics, one concentrates on Kinematics and the other in Dynamics. Nevertheless, both exercises have a common feature: a mechanism on which questions can be posed. The next is a list of considerations taken into account to design a useful exercise:

1. The mechanism object of the exercise must include at least one of each type of the kinematic pairs (joints) studied during the course.
2. The configuration of the mechanism, position of solids, must be different for each student. This is usually achieved by parametrization of lengths and distances in terms of the ID number of the student. In order to minimize the possibility of two students having the same configuration, an irrational number in the range (0, 1) is defined as follows:

$$\xi = \frac{\sqrt[8]{\prod_{i=1}^{8} A_i}}{2 + \frac{1}{8}\sum_{i=1}^{8} A_i} \tag{1}$$

   where A_i are the eight digits of the student ID number assuming that a 0 means 10 to avoid singularities.
3. Singular configurations (null Jacobian) must be avoided since the solution code would give NaN results which can not be compared with those of the student. This may be difficult to achieve in all cases and is highly dependent on the experience of the teacher. The experience of the authors reveals that the use of the number ξ helps to avoid singular configurations when it is used in the following way. The mechanism topology is divided in angular sectors that contains the geometric vectors involved in the calculus of velocities and accelerations. Then, the orientation of these vectors is taken as a weighted mean between the limits of the angular sectors. As an example, the radius vector of a point A could be calculated as

$$[r]^A = [r]^{C_5} + R_6 \begin{pmatrix} \cos\gamma \\ \sin\gamma \end{pmatrix} \tag{2}$$

   where $[r]^{C_5}$ is known, R_6 is parameterized in terms of the ID digits and $\gamma = \frac{2\pi}{3} + \xi \cdot \frac{\pi}{6}$ rad. Note that in this case the limits of the angular sector are

$\frac{2\pi}{3}$ and $\frac{5\pi}{6}$. Controlling the orientation of vectors in this way it is possible to avoid singular configuration in almost any case.
4. Several inputs and outputs must be available in order to generate combinations of inputs and an output that increase the number of different exercises. In general, the mechanisms used in this kind of exercises would be single d.o.f. mechanisms. Therefore it is useful to parameterize also which are the inputs and which is the output.
5. Finally, it is necessary to pose questions that can be answered numerically. The numerical solution of this kind of exercise involve many computations and therefore it is recommended that the student codes the resolution of the process. In despite of this recommendation many students take the risk of solving the problem manually, which is an excellent training for the written exam.

After grading all students, Doctus application allows the teacher to make visible the results corresponding to each students. This way, even if the grade got by a student was not good he/she still can learn from repeating the exercise until the exact solution is achieved, a process that may point to possible sources of mistakes.

3 Exercises Proposed During Course 2015–2016

Figures 1 and 2 shows the two mechanisms that where object of kinematic and dynamic analysis during the course 2015–2016[1]. In the figures, one can find all the parameters that define the configuration of the mechanism. For instance, parameter R_6 in Fig. 1 is defined as A_2 meters, being A_2 the second digit of the ID number.

To show how the input data can be parameterized, in the kinematic analyses of the mechanism in Fig. 1 the value of the independent velocity was given according to

- $v_{32} = 1$ m/s if A_7 is 1, 2 or 3.
- $\omega_{31} = 1$ rad/s if A_7 is 4, 5 or 6.
- $\omega_{41} = 1$ rad/s if A_7 is 7, 8, 9 or 10.

being A_7 the seventh digit of the ID number.

The mechanism in Fig. 2 was subject to the dynamic analysis. Again, the problem is completely parameterized including forces or moment applied to different bodies. In this case, the following types of analysis were requested on the same configuration (position) of the mechanism: a static equilibrium scenario, an inverse dynamic analysis with the mechanism in repose (null velocities), a direct dynamic analysis with the mechanism in repose, an inverse dynamic analysis with the mechanism in movement and a direct dynamic analysis with the mechanism in movement. The method suggested to solve the problem is the

[1] The complete exercise is available on request to dgvallejo@us.es.

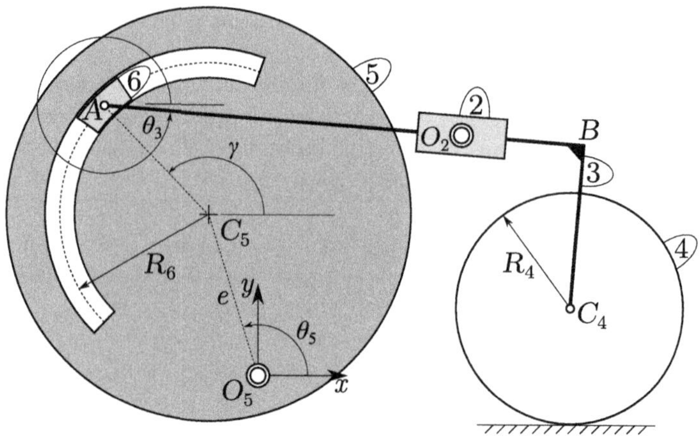

Fig. 1. Draft of the mechanism whose kinematics has to be analyzed.

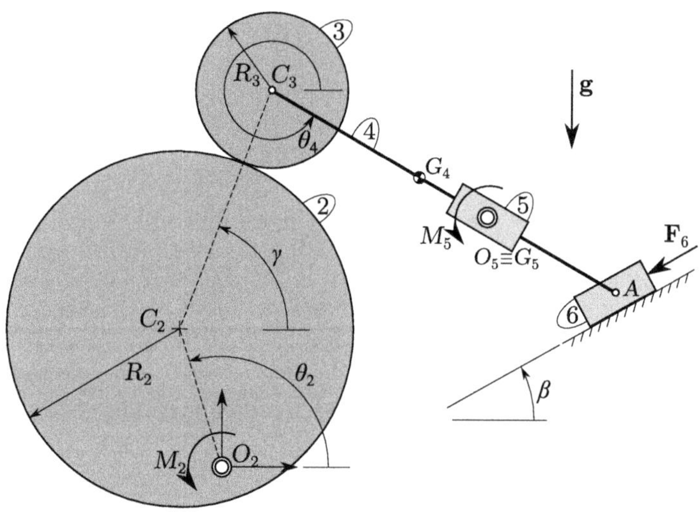

Fig. 2. Draft of the mechanism whose dynamics has to be analyzed.

application of the Virtual Power Principle. This five previous cases helps at clarifying which situations require null inertial forces or when virtual velocities need to be different to real ones, for example.

The kinematic analysis was carried out by 393 out of 462 students. Among them 123 obtained the highest grade while 77 obtained the lowest. The mean grade was 6.23 over 10. On the other hand, the dynamic analysis exercise was done by 344 students. Only one student got the highest grade while 171 student got the lowest grade. The mean grade in this exercise was 1.81 over 10.

4 Correlation Between Exercises and Exam Grades

It is also important to analyze whether the exercises had any effect in the students' learning achievements. To do so the grades obtained by the students in final exam was correlated with those obtained in the exercises during the academic year 2015–16. That year 462 students were enrolled in the course, though only the 96 students attending the exam in June were considered for the study. Their grades in the part of the exam devoted to Kinematics and Dynamics of Mechanisms (named EXAM) were correlated with the corresponding grades obtained in the two exercises proposed during the year and explained in Sect. 3. These grades are named, respectively, KIN and DYN.

It must be noted that EXAM was graded in the range $[0, 10]$ and using quarters of point as the grading unit. Meanwhile, KIN and DYN were graded, respectively, in the ranges $[0, 6]$ and $[0, 9]$, being 1 the grading unit. In other words, KIN and DYN had, respectively, 6 and 9 questions and only true or false (1 or 0) were accepted as grades in each question.

Pairwise comparisons of the three grades can be seen in Fig. 3.

First, a Kolmogorov-Smirnov test using the Lilliefors significance correction was performed to check the normality of the samples. The test was significant in all cases ($p < .001$ for KIN and DYN and $p = .007$ for EXAM) and, thus, the correlation performed between the grades was based on the ranks.

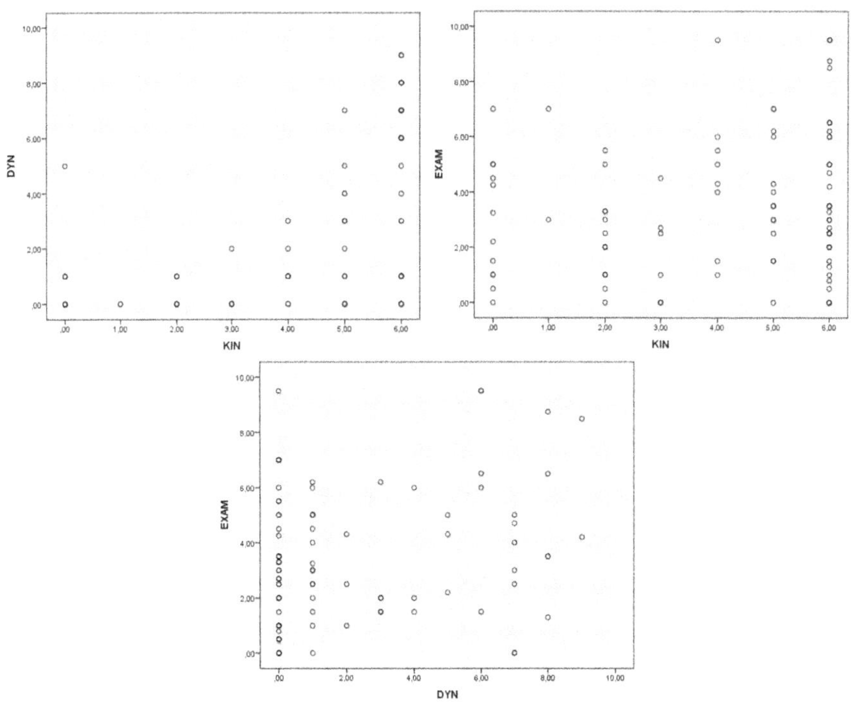

Fig. 3. Pairwise comparisons of the grades EXAM, KIN and DYN.

The grade EXAM was significantly correlated with DYN (Spearman $R = .217$ and $p = .034$) but not with KIN (Spearman $R = .152$ and $p = .138$). In the first case $R^2 = .047$ is not very high, but it is obvious that the students' grades in a certain exam may depend on multiple variables. That year the grade obtained in the Dynamics exercise explained almost 5% of the grade in the final exam and that is not negligible. On the contrary, it is revealing that a continuous study of the course helped the students to achieve good results.

One of those multiple effects that may affect the grades of an exam is the difficulty of the exercises, which did not seem high in case KIN. This could explain the low correlation between KIN and EXAM. However, it is noteworthy the strong correlation found between KIN and DYN (Spearman $= .508$ and $p < .001$). Here, the similar conditions upon which the exercises were completed by the students: at home, with time enough to solve the problems, with the help of their textbooks, etc. may be behind that stronger correlation.

Another factor that is surely biasing the comparisons is the fact that the grades KIN and DYN were calculated using a true/false scheme, based on the results and an acceptable tolerance, in contrast to EXAM in which the process followed to solve the problem was taken into account to calculate the grade.

Acknowledgements. The free access to the application and the continuous help and support provided to the authors of this paper by the leaders of Doctus project, David Muñoz de la Peña and Fabio Gómez-Estern, is sincerely acknowledged.

References

1. Madrigal Sánchez, C., del Pozo Polidoro, E., Chaves Repiso, V., Chamorro Moreno, R., Martínez Reina, F.J., García Vallejo, D., Escalona Franco, J.L., Ojeda Granja, J., Navarro Pintado, C., Domínguez Abascal, J., Vázquez Valeo, J., Reina Romo, E.M., Acosta Muñoz, M.: Teoría de Máquinas y Mecanismos. Editorial de la Universidad de Sevilla. ISBN: 978-84-472-1856-1
2. The MathWorks, Inc.: MATLAB. 1994–2017. https://es.mathworks.com/products/matlab.html
3. Sánchez Cazorla, C., Muñoz de la Peña Sequedo, D.: Evaluación automática para problemas de control. Universidad de Sevilla, Tesis Doctoral (2015)

Application of 3D Printing to the Intersection of Surfaces Learning in a Descriptive Geometry Course

A. M. Gómez-Amador(✉), L. Prieto-Fernández, E. Olmeda, and V. Díaz

Mechanical Engineering Department, Universidad Carlos III de Madrid, Avda. De la Universidad, nº30, 28911 Leganés, Spain
amgomez@ing.uc3m.es

Abstract. This work on surfaces intersection intends to facilitate learning on this subject to the student, something that, from teaching experience, has been found to be difficult to understand and to visualize.

The student's ability to resolve or discriminate spatially defined features is a very important requirement when it comes to understanding certain types of intersections of surfaces, due to its wide application in the engineering framework.

This work shows that the use of 3D printed models facilitates learning the dihedral system and improves the performance of the student on problem solving.

Keywords: Descriptive geometry · Surface intersections · 3D printed models

1 Introduction

The Descriptive Geometry studies include concepts such as abatements, change in the projection plane, parallelism and perpendicularity, shadows, etc., but without doubt the subject that requires the most from the student's spatial vision is surface intersections.

The complexity of this subject, as well as the limited time available for its development in the classrooms, makes understanding difficult for the student. The introduction of new and innovative teaching methods demonstrates that the student's spatial understanding improves and increases by optimizing the learning process [2].

With the audiovisual and the traditional media, like the blackboard, the professor can explain the general procedures to determinate the surface intersection and how to obtain the curve limiting points at the top, bottom, left and right side.

The intersection of two surfaces is obtained by using an auxiliary surface that intersects them and obtaining the sections that it produces in each of them. The common points of these sections belong to the intersection line of the two surfaces. Repeating this process and joining the points, we obtain the intersection curve between both surfaces. This process will be repeated as many times as necessary so that the union of the points of intersection is clear and does not lead to any error.

J. C. García-Prada and C. Castejón (Eds.): *New Trends in Educational Activity in the Field of Mechanism and Machine Theory*, MMS 64, pp. 165–173, 2019.
https://doi.org/10.1007/978-3-030-00108-7_19

The choice of the auxiliary surfaces will depend on the type and position of the surfaces which are object of the problem. Generally it can be indicated that the auxiliary surfaces will be planes or spheres. The intersections of quadrics are those that occur most frequently. The intersection of two quadrics can be a closed curve, named "bite", or two curves independent of each other, one the entrance and the other the exit hole, named "penetration". In general, the intersection curve of two quadric surfaces is a twisted curve [1].

With the application of traditional methods, like the blackboard and audiovisuals, the students identify the main surfaces of the piece that they are modelling and they can obtain the manufacturing drawings with the linear dimensions but, in general, they have many difficulties to draw the intersections. Computer aided drawing software can facilitate the visualization of intersections, but it does not indicate to the student how to obtain them. Therefore, we have concluded that it is extremely useful the use of 3D pieces to achieve complete and fast learning skills based on the principle of "touch in order to see".

At the beginnings the 3D pieces were manufactured in cardboard material, but its fragility precluded its repeated use. This problem was solved with pieces created by the 3D printer.

2 Intersection of Surfaces

The design and draughting of the manufacturing drawings of a mechanism requires that the student can assemble the pieces that constitute it. The visualization of the different types of intersections, which occur between the different parts of the mechanism, is important for the choice of the type of material, the manufacturing processes, etc. The engineering student must provide viable technology proposals and optimized solutions.

Optimized learning is achieved when real applications can be used to understand the intersection curves. For example, if the student can associate the intersection of two cylinders to a T-connection pipe, the visualization of the problem is faster and the execution of the drawings easier.

Following is shown one of the examples of mechanical pieces used in the classroom to explain the effect of the intersection surfaces on drawings.

The piece of Fig. 1 represents a tool that serves to drive a device. The tool allows two forms of driving, both manuals; the first is to rotate the tool axially using the four cylindrical bars that appear at the bottom; the second allows coupling a bar in the central hole of the tool to increase the drive torque to the device.

The coupling of the tool with the device is carried out in the upper part, in the area of the inclined lifting holes, by fixing screws not shown in Fig. 1.

Geometrically the tool consists of a squared prism at the bottom zone, a truncated cone in the central zone, and a cylinder in the upper zone of the part piece. Four rounded rods come out perpendicularly from the prism. The upper cylinder is cut by two planes at 45°; these planes serve as the base for the inclined lifting holes. The cylinder also has a central hole with a dovetail shape for the assembly of the drive rod.

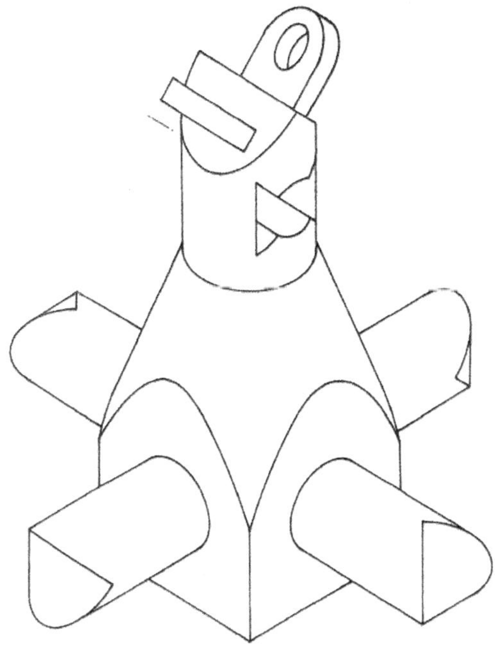

Fig. 1. Manually operated tool

The intersections of surfaces present in this tool are the following:

- Intersection of the central truncated cone with the prism. The prism surfaces are parallel planes to the cone axis, so the intersection curve is defined by the four hyperbola branches generated by each of the faces of the prism.
- Rounding the end of the four bottom rods. This curve is an intersection of two cylinders with the same diameter and perpendicular axes. One of the cylinders is the solid rod and the other cylinder is to round out the external edge. The intersection is a "bite" case with two double points. In the isometric projection one of the branches of the intersection projects according to a straight line in the direction of the coordinate axes. To obtain the intersection, it is enough to cut the two surfaces by horizontal planes, which will provide, as an intersection with the two cylinders, circumferences in one and generatrix in the other.
- Intersection of the cylinder at the top zone and oblique planes. These are two ellipse branches, one of which is projected according to a line in the isometric projection.
- Intersection of the cylinder at the top zone and the central hole with dovetail shape. This intersection is a "penetration" case with an entrance curve and an exit curve. The hole is formed by the three planes that define the dovetail and a cylinder with horizontal axis.

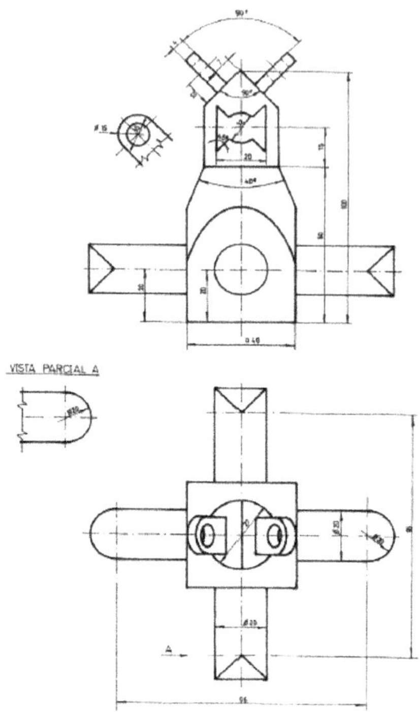

Fig. 2. Tool manufacturing drawing

Understanding all the intersections of Fig. 1 allows the student to develop the manufacturing drawings shown in Fig. 2. It includes a main view, named front view, and a top view. In both views, the intersections described above appear. It is interesting that the student understands that it is not necessary to give dimensions to the intersections, only dimensioning of the initial surfaces is required. For instance, the hyperbolas in the prism are not dimensioned, since they are a direct consequence of the truncated cone and the prism dimensions. That is, the student learns to distinguish which zones of the views are intersections, to avoid their dimensioning, and which zones of the views correspond to the projection of the initial surfaces and have to be dimensioned.

3 Use of 3D Printed Pieces

The student does not always have the isometric perspectives, like the one shown in Fig. 1, and must be able to solve exercises of views with intersections, such as the one shown in Fig. 3.

Figure 4 shows the resolution in dihedral system of an intersection between a cone and a cylinder. Its resolution requires a complex change of the projection plane and the use of auxiliary spheres. In these cases, the student finds serious difficulties to solve the exercise and find the solution shown in Fig. 4. In these exercises the students are aware of the need to use auxiliary elements, but it is not easy for them to visualize the problem.

The computer-aided drawing software provides the student with the problem solution but it does not show the method used to obtain it.

For a correct interpretation of the problem, cardboard models of pieces with intersections were introduced in the class as teaching material. This way to visualize the intersections turned out to be relevant, but the models deteriorated easily due to the low resistance of the material and the manipulation by students.

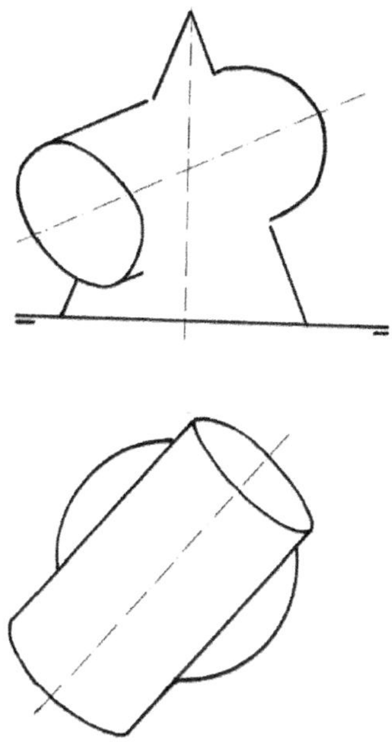

Fig. 3. Dihedral system: cone and cylinder intersection

The results obtained with the cardboard prototypes were good and it was decided to try again with prototypes printed in 3D. These models of easy design and made of a material resistant to manipulation by students are of utmost importance as a complement to the theoretical classes. They allow to visualize the problem and "touch" it, in order to make easy the drafting.

In the first phase, two pieces have been printed as prototypes [3]:

- Two cylinders of equal diameter and perpendicular axes.
- Two cylinders of different diameter and perpendicular axes (Fig. 5).

To evaluate the results of the use of 3D printed pieces as teaching material, the student is provided with the isometric perspectives of cylinders of equal and different diameter shown in Figs. 6 and 7 respectively.

Later, the same pieces are provided in the dihedral system and it is observed that the students do not understand the drawings although they identify the intersections based on the theoretical classes. However, with the distribution of 3D printed prototypes (Figs. 8 and 9) the student fully understands the given views.

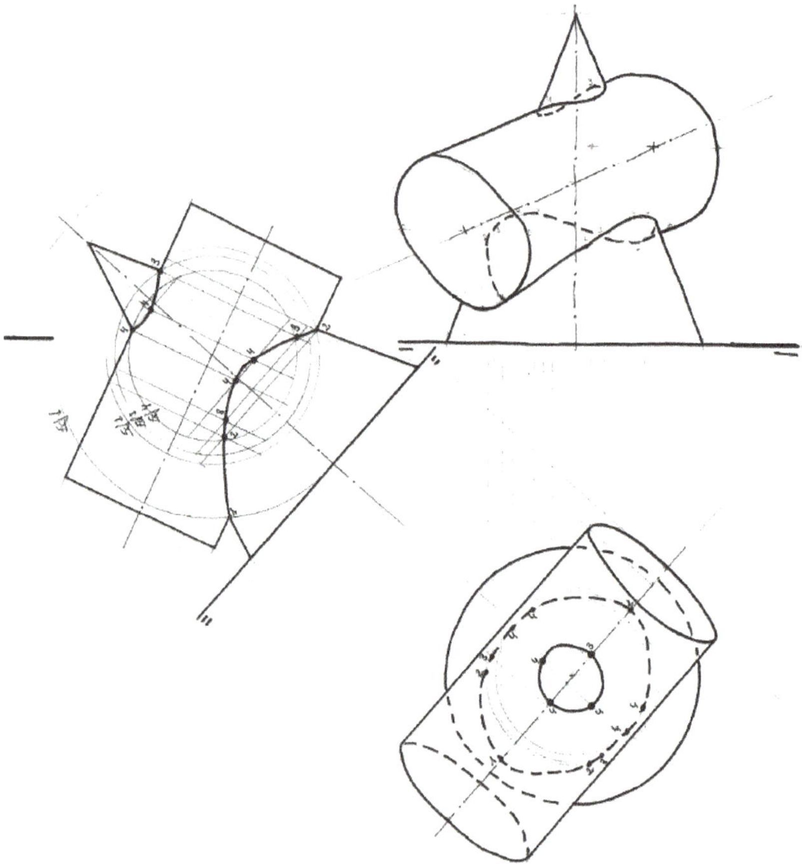

Fig. 4. Dihedral system: cone and cylinder intersection solution

Fig. 5. 3D printer

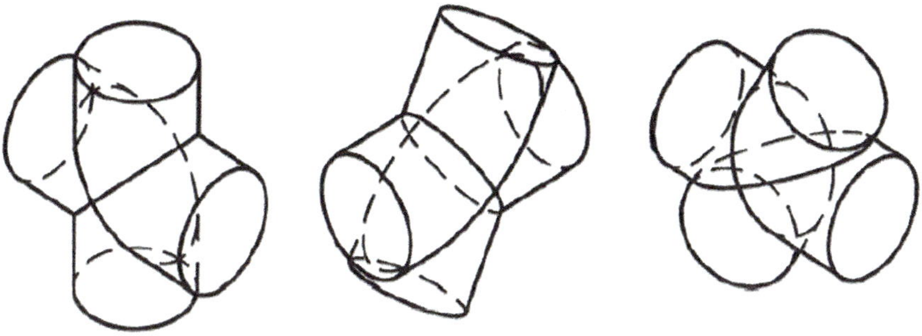

Fig. 6. Isometric perspectives: intersection of same diameter cylinders

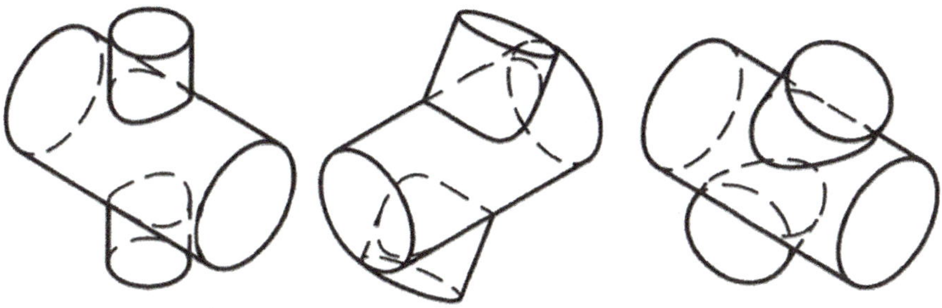

Fig. 7. Isometric perspectives: intersection of different diameter cylinders

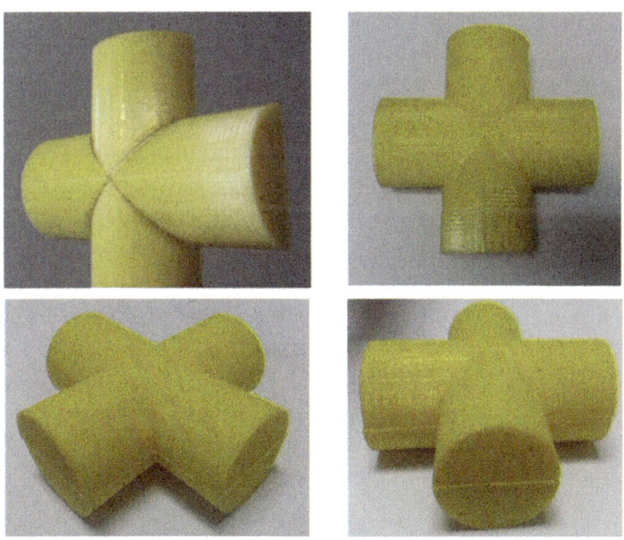

Fig. 8. 3D printed piece: intersection of cylinders of equal diameter

Fig. 9. 3D printed piece: intersection of cylinders of different diameter

4 Conclusions and Future Works

Based on the accumulated teaching experience in the Descriptive Geometry, it has been found out that students need help in the resolution of surface intersections. This article is based on that experience of several years and its conclusions are intended to be implemented in future courses.

The use of 3D printed pieces as teaching material would significantly help the student to understand the problem and its resolution.

Problems of difficult visualization can be solved when the student can see and touch the pieces. Prototypes used for cylinder intersections have been found very useful for the correct understanding of the views, so in the future the models will be extended to other geometries that appear on a regular basis in the industrial designs.

Any tool to optimize the teaching effectiveness is an essential element, mainly when there is a reduction in the number of the classes dedicated in an engineering course to the descriptive geometry and the isometric perspective.

Additionally, the involvement of the students in the training process is enhanced, facilitating their participation, improving their spatial vision and capturing their attention constantly in the classroom. All these factors lead to a greater achievement of the teaching objectives proposed in the subject.

References

1. Javier Rodríguez de Abajo, F.: Geometría Descriptiva: Sistema Diédrico. Editorial Donostiarra (2007). ISBN: 9788470633539
2. OECD: Innovating Education and Educating for Innovation. OECD Publishing, Paris (2016)
3. Madrid Vicente, A.: Tecnología de la impresión 3D. Editorial AMV Ediciones (2017). ISBN: 978-84-945558-2-2

Mechanical Engineering Dducation: Experiences

Didactic Methodologies Used in Industrial Design and Mechanical Engineering for the Implementation of the Marked Competencies and Their Professional Insertion

J. M. García-Alonso[1(✉)], E. Soriano-Heras[2], F. Blaya[3], and H. Rubio[2]

[1] Salesianos Atocha, Madrid, Spain
jesusmanuel.garcia@salesianosatocha.es
[2] Universidad Carlos III de Madrid, Madrid, Spain
{esoriano,hrubio}@ing.uc3m.es
[3] Universidad Politécnica de Madrid, Madrid, Spain
fernando.blaya@upm.es

Abstract. Since the academic year 2010–11 the Technical University of Madrid (UPM), University Carlos III of Madrid (UC3M) and the Salesianos Atocha College (CSA), have been steadil working on teaching their subjects through the Project-Oriented Learning (POL) methodology. Later they also implemented more didactics methodologies in order to achieve a higher development in the competences of students. These schools incorporated Service-Learning (SL), and several other tools such as Dynamic Technical Documentation (DTD), Google Sites and Moodle. Other teaching methodologies have been added since the academic year 2014–15 in order to further improve the training of students, which pay special attention to training units which are particularly difficult to understand. To this end, didactic methodologies such as Kounaikenshuu, Flipped Classroom, and tools such as Portfolio, Plickers and Kahoot! all aimed at improving the quality of the Bachelor Final Thesis. The implementation of these new didactic methodologies not only improved students' academic training (32% improvement in UPM, 14% in theoretical training and 33% in practical training in CSA) but also achieved the competencies marked for the double Degree in Design and Mechanical Engineering at UPM and Industrial Mechatronics in the CSA These results were also confirmed by the satisfaction surveys carried out by the collaborating partners of the projects which showed improvements of between 19% and 43% in theoretical training and between 53% and 68% in practical training at UPM, whereas in CSA, the improvement in theoretical training was 22% and 85% in practical training according to the criterion of the partner-collaborators. Professors and teachers from both institutions said that the implementation of these methodologies caused them to be overworked due to the excessive number of students but on the other hand, they also improved their teaching competencies.

Keywords: Kounaikenshuu methodology · Flipped-Classroom Service-Learning · Project-Oriented Learning · Collaborative work

J. C. García-Prada and C. Castejón (Eds.): *New Trends in Educational Activity in the Field of Mechanism and Machine Theory*, MMS 64, pp. 177–185, 2019.
https://doi.org/10.1007/978-3-030-00108-7_20

1 Introduction

For an educational innovation to be considered as such, it must fulfill two requirements: "Educational innovation is the application of an idea that produces planned change in processes, services or products that generate improvement in training objectives" and, on the other hand, which has a series of characteristics: originality, efficiency, efficiency, transferability and sustainability. So, not all the proposals that are usually presented as educational innovations are such.

Therefore, for an experience to be considered an educational innovation, it must be verified that it satisfies the following principles [1]:

1. Originality. It should consider previous experiences that have been made in the field of training. The originality is not usually given by the applied technology, but by the way of using that technology to improve the methodologies.
2. Effectiveness. It must meet the objectives for which the innovation was designed and demonstrate that they are met the scientific method.
3. Efficiency. The cost/effort to apply educational innovation should never exceed the cost/effort made before applying it.
4. Transferability. It should be able to be used in other subjects.
5. Sustainability. Once the educational innovation is developed its application does not require new developments. For example, many experiences of educational innovation are realized because there is funding and once this is completed, they continue to require economic contributions. In this case, innovation would not be sustainable.

Innovations that are taking place at a fast pace at all levels; technological, labor, social, etc., require versatile and flexible engineers, capable of adapting quickly to the needs of the industrial world where the "partner-productive" model has been imposed, universities are required to change their teaching-learning models and adapt to the new challenges. To this end, traditional didactic methodologies must be replaced by active didactic methodologies which modify the learning of theoretical and practical concepts, as well as aspects of planning, realization, thinking,… enhancing interdisciplinary, social and cultural competencies.

Companies like HP, Telefonica, Ascent, Seat, Mastercard, Educaweb or Humantiks powered by the Mobile World Congress (MWC) are trying to bring STEAM (science, technology, engineering, art, and mathematics) careers closer to elementary students through the use of leisure activities prepared for this purpose This is because it is predicted that in a few years 65% of elementary students will choose careers that will not be relevant and, on the other hand, there are fewer and fewer students choosing STEAM subjects, especially women, with a growth of 14% in Europe and 10% in Spain by the year 2020. These profiles are currently falling, with 25% in Europe and 40% in Spain.

Due to all of this, in view of the future needs and the rapid changes that are occurring in society, it is important to establish in education an initiative, to go one step beyond what is explained in the classroom, which emphasizes researching, experimenting and analyzing, as well as seeking solutions to the problems that exist today in society. An example of this could be the resolution of the ciber-attack problem which occurred on May 12, 2017 around the world, and which was partially solved by a 22-year-old UK Computer Engineering student, Marcus Hutchins.

Scientists and engineers are now generating technological advances at an accelerating pace. To be able to produce this, other solid theoretical knowledge, skills or abilities such as teamwork, entrepreneurship, verbal or written communication, curiosity, creativity and, above all, a proactive attitude are required [2]. In this line, several authors have stated that for a technological development the role of engineering is fundamental [3–5].

Currently, there are certain studies around the world, which are the most appropriate methodologies for learning for engineering students. Some authors have expressed that the use of information and communication technologies will be fundamental in near future [6–8]. Other authors, on the other hand, have focused attention on pedagogical models, educational curricula, curricula, etc., fostering skills such as creativity, teamwork,… [9–12].

In the present article, a study is presented on the didactic methodologies used at UPM, in fourth year of double Degree in Industrial Design and Mechanical Engineering, in the subjects of: Technical Office, Mechanical Design, Three Dimensional Digitization and Rapid Prototyping and Analysis and Synthesis of Mechanisms, and within the participation of CSA, specialization in Industrial Mechatronics, in the subjects of: Management Processes of Computer Aided Maintenance (CMMS), Configuration of Mechatronic Systems, System Integration and, Simulation of Systems.

For more information on the foundations used at UPM, in the double Degree in Engineering in Industrial Design, the article published by this same Department for the Congress ICERI 2016 [13] can be consulted.

2 Methodology

Until 2014, the POL methodology was used at UPM and CSA, where a team of professors presented students with real projects to put into practice the theoretical contents taught in the subjects, producing students who were more proactive, responsible and more committed to their learning. Within this learning methodology, students achieved an interdisciplinary and global vision of all the participating subjects, while teachers improved in aspects such as teamwork, relationships with other departments [14].

Noting that there were improvements, but that certain important deficiencies had not yet been solved, both in theoretical concepts and in some of the skills to be developed, the team of teachers decided to add new didactic methodologies, with the idea of improving the formation of the future engineers, as well as the high degree technicians in vocational. This new strategy of methodologies is shown in Table 1. As can be seen, these methodologies are not only focused on students, but also on the teaching team. Thus, for example, the Kounaikenshuu (K) methodology helps the teaching team to improve the documentation of each of the subjects, beginning with those units that are harder for students to understand.

Table 1. Subjects participating in the experience

Didactic methodologies	Made by	Subjects	Hours	Month
Kounaikenshuu:	Teaching staff	Technical office	6	June to September
- Yugyou Kenkyuu		Mechanical design.	4	
- Yugyou Bunseki		Analysis and synthesis.	4	
Flipped Classroom:		Digitization	4	
Software:	Teaching staff	Management processes	6	
- Plickers		System settings	7	
- Kahoot!		System integration	7	
- Moodle		Systems simulation	4	
Project-Oriented Learning (POL), through Service-Learning (SL) projects.	Students UPM-CSA	Technical office		September to January
Flipped Classroom (FC).		Mechanical design		
Collaborative Work (CW).		Analysis and synthesis.		
Use tools:		Digitization		
- Google Drive				
- Plickers				
- Kahoots				
- Moodle				

As for the methodologies applied to students, it should be said that the POL was a great success when it was implemented in the double Degree in Design Mechanical Engineering, with the help of the contents uploaded to Google Sites, accompanied by videos created by the teaching team, so that students could use them, at any time or location, These projects are the epicenter of this initiative? and the professors are the facilitators and companions of the students' learning [15].

The collaborative work has been carried out through the platform Google Sites, where students consult, take notes, record in a portfolio what they are doing on a day to day basis, and work together on the common document of the project they are working on.

In some subjects, programs such as: Plickers and Kahoot! have been used from time to time to obtain feedback on the comprehension of the subject.

As shown in Table 2, in sessions of 2 and a half hours, it was possible to devote to the problems or the project up to 85 min, being very positive for the student the use of classroom hours. Nevertheless, students should perform outside the classroom (home, library, etc.) personal work of about 40 to 50 min per day. This is the time in which the student develops a deeper understanding of the subject, being the responsibility of each student, the discussion and resolution of those problems that in traditional teaching were raised during sessions time.

Table 2. Temporal distribution for UD per week

Outside the classroom	Classroom			Laboratory	
Personal work	Introd.	Individual test	Test equipment	Rating/correction	Hours week/project time
40' a 50'	Max 10'	Max. 20'	Max. 20'	Max. 15'	2 h. 55'
					2,5 h. 85'
					4 h. 175'
					4,5 h. 205'

The experience has taken place in six phases. An extensive development of the process carried out during the academic year is shown in Table 3.

Table 3. Phases followed in the realization of the projects, both professors and students

Month	Week	Professors	Project phases	Students
May–August		Presentation		
September	First			
	Second	Presentation	Start project	Creation of working groups. Project Planning
October	Fourth	1st deliverable	1st delivery portfolio Initial design sketches	Delivery and 1st presentation what has been done
	First	Assessment of classroom attitudes	Reference searching	
	Second	Assessment of classroom attitudes	Reference searching	
	Third	2nd deliverable	2nd delivery portfolio. More advanced calculations and design	2nd Presentation
	Fourth	Assessment of classroom attitudes	Communication, leadership, etc	

(continued)

Table 3. (*continued*)

Month	Week	Professors	Project phases	Students
November	First	3rd deliverable	3rd delivery portfolio. Plans and Budget	
	Second	Assessment of classroom attitudes		3rd Presentation (75% done)
	Third	4th deliverable	4th delivery portfolio. Specification and maintenance	
	Fourth	Assessment of classroom attitudes		
December	First	5th deliverable	5th delivery portfolio. Start 3D Printer Assembly	
	Second	Assessment of classroom attitudes		4th Presentation (90% done)
	Third	6th deliverable	6th delivery portfolio. Finished 3D printer	
January	Second	Final delivery		
	Third			Project final Presentation, critical self-evaluation

The evaluation of each subject, both at UPM and at C.S.A., consisted of: A Theoretical Exam (TE), a Project (P) and student's degree of Interest (I). The Final Assessment (FC) was obtained after applying the formula showed in the next equation.

$$\mathrm{FC} = 0.3 * \mathrm{TE} + 0.65 * \mathrm{P} + 0.05 * \mathrm{I}$$

3 Results and Analysis

From the 2014–15 academic year, the methodologies described in the previous section were implemented. The assessments obtained by the students from the double Degree in Industrial Design and in Mechanical Engineering, in the subjects of: Technical Office, Mechanical Design, Three-Dimensional Digitization and Analysis and Synthesis of Mechanisms are shown in Fig. 1. The assessments obtained from the application of the active methodologies applied since the academic year 2014–15 have shown better learning and understanding of theoretical concepts.

Not only have the results of theoretical knowledge been better achieved, but also certain skills have been developed such as ideas, analysis, presentations, software management. For example, in this academic year the average assessments obtained in

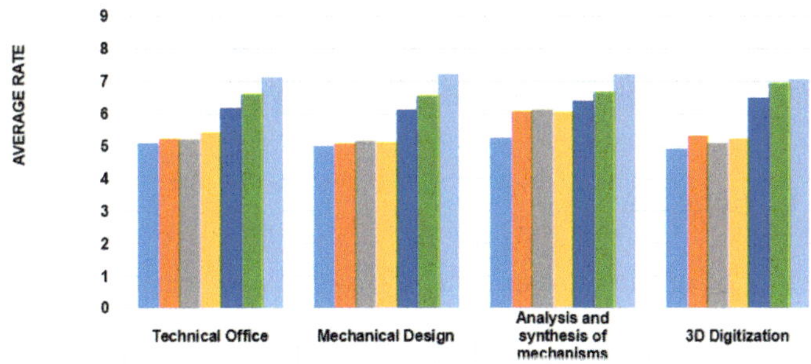

DEGREE IN ENGINEERING IN INDUSTRIAL DESIGN	Technical Office	Mechanical Design	Analysis and synthesis of mechanisms	3D Digitization
WITHOUT APPLICATION OF: POL-SL-FC-Koun ... Course 2010-11	5,1	5	5,23	4,9
WITHOUT APPLICATION OF: POL-SL-FC-Koun ... Course 2011-12	5,2	5,05	6,07	5,3
WITHOUT APPLICATION OF: POL-SL-FC-Koun ... Course 2012-13	5,2	5,16	6,14	5,1
WITHOUT APPLICATION OF: POL-SL-FC-Koun ... Course 2013-14	5,4	5,08	6,04	5,2
WITH APPLICATION OF: POL-SL-FC-Koun ... Course 2014-15	6,2	6,14	6,41	6,5
WITH APPLICATION OF: POL-SL-FC-Koun ... Course 2015-16	6,6	6,55	6,68	6,93
WITH APPLICATION OF: POL-SL-FC-Koun ... Course 2016-17	7,14	7,23	7,23	7,08

Fig. 1. Mean ratings UPM. Theoretical knowledge

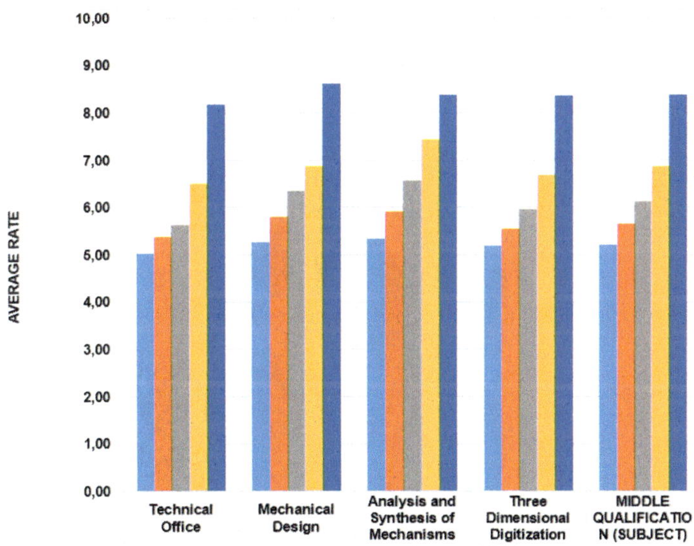

DEGREE IN ENGINEERING IN INDUSTRIAL DESIGN	Technical Office	Mechanical Design	Analysis and Synthesis of Mechanisms	Three Dimensional Digitization	MIDDLE QUALIFICATIO N (SUBJECT)
FIRST PRESENTATION	5,03	5,26	5,34	5,19	5,21
SECOND PRESENTATION	5,37	5,79	5,91	5,55	5,66
THIRD PRESENTATION	5,63	6,34	6,57	5,97	6,13
FOURTH PRESENTATION	6,50	6,86	7,42	6,67	6,86
FIFTH PRESENTATION (FINAL)	8,17	8,62	8,37	8,36	8,38

Fig. 2. UPM projects presentations course 16/17

presentations of the projects at UPM in each of the subjects, are shown in Fig. 2. computer programs such as MS PowerPoint, Prezi or Emaze were used during presentations in similar ways there were improvements at CSA.

4 Conclusions

The quantifiable improvements in the students participating in the initiative have been:

- Improvements in the final assessments at UPM. They have varied between 19% of Analysis and Synthesis of Mechanisms to 43.16% in Mechanical Design, and at C. S.A., they have varied between 14.66% in Integration of Systems to 62.26% in Configuration of Mechatronic Systems.
- From both institutions, were reduced to between 3 to 6%, due to health or financial problems.
- Due to all of the above, the improvements in the final assessments were between 27 to 36.5% with respect to the academic year 2013–14 where the active methodologies were started.
- A novel aspect in this academic year was a project involving augmented reality, see Fig. 3.

Fig. 3. Augmented reality. Cartesian Robot

References

1. Sein-Echaluce, M.L., Fidalgo-Blanco, A., Alves, A.Y.: Technology behaviors in education innovation. Comput. Hum. Behav. **72**, 596–598 (2016)
2. Tippelt, R., Lindemann, H.: El método de proyectos. El Salvador (2001)
3. Bush, V.: Science: the endless frontier. Trans. Kans. Acad. Sci. **48**(3), 231–264 (1945)

4. Keating, D., Stanford, T.G.: An innovative strategy to integrate relevant graduate professional education for engineers in industry with continual technological innovation. In: SEE Annual Conference Proceedings (1999)
5. Ciampi, M.M., Da Brito, C.R.: Awareness of social impact of engineering: the task for engineering schools. American Society for Engineering Education (2011)
6. Da Brito, C.R., Ciampi, M.M.: The discussions after the Bologna process in Europe: the global engineer. American Society for Engineering Education (2011)
7. Tao, Y.P., Wang, X.B., Li, K.Q., Liang, Z.: Assessing Chinese engineering graduates' abilities for problem-solving, scientific discovery and technological innovation from a professoriate perspective. Age **5**, 1 (2000)
8. Seery, N., Donal C., Dunbar, R.J.: Maximising the impact of creative and innovative activities within the constraints of defined educational structures. In: ASEE Conference and Exposition. American Society for Engineering Education, Louisville (2010)
9. Islam, R.: Engineering and technology education in Bangladesh: comparative study of the public and private universities for problems and prospects. In: Conference Proceedings of the ASEE Annual Conference and Exposition (2012)
10. Vaezi-Nejad, S.M., Cullinan, M., Bishop, P.: Telematics education I: teaching, learning and assessment at postgraduate level. Int. J. Electr. Eng. Educ. **42**(2), 132–146 (2005)
11. Godfrey, D.: A new approach to teaching introduction to electrical engineering at the United States coast guard academy. In: ASEE Annual Conference and Exposition (2006)
12. Zheng, W., Huiru S., Mo, Y.-L.: Integration of cognitive instructions and problem/project-based learning into civil engineering curriculum to cultivate creativity and self-directed learning skills. American Society for Engineering Education (2009)
13. Blaya, F., Nuere, S., Soriano, E., García, J.M., Islán, M. Implementation of a methodology project oriented learning (POL) on specific subjects of bachelor and master in industrial design. In: ICERI Annual Conference Proceedings (2016)
14. Garcia, J.M., Soriano, E., Garcia, I., Rubio, H.: Implementation of service-learning projects in engineering colleges. Int. J. Electr. Eng. Educ. **29**(5), 1119–1125 (2013)
15. de Miguel Díaz, M.: Modalidades de enseñanza centradas en el desarrollo de competencias. Orientaciones para promover el cambio metodológico en el Espacio Europeo de Educación Superior (2005)

Educational Resources for Self-learning of Descriptive Geometry

C. Álvarez-Caldas(✉), J. Meneses, S. Santos, A. M. Gómez, A. Díaz, J. A. Calvo, and A. Muñoz

Mechanical Engineering Department, Universidad Carlos III de Madrid, Avda. De la Universidad, nº30, 28911 Leganés, Spain
calvarez@ing.uc3m.es

Abstract. In this work, two educational resources for self-learning of Descriptive Geometry are presented: the "Zero Course" and the "Support Course". The creation of this e-learning material responds to the need that our students at University Carlos III de Madrid have to reach a minimum level at the beginning (Zero Course), and during (Support Course) the first course on technical drawing. First, the need is made out through results of surveys carried out in previous years. Then, some e-learning applications are exposed, among which the most appropriate to the need are chosen. Finally, the designed courses are described including all the technical resources. The results of the surveys carried out on the students of these courses, as well as some statistics of their qualifications, are also presented.

1 Introduction

Descriptive Geometry (DG) is the branch of Geometry that studies the representation of three-dimensional objects on a plane, using systems based on the concept of projecting the object on a plane in order to reduce the three spatial dimensions to the two dimensions of the plane. At present, the basic content of DG (as the basis of the Orthographic Projection) is taught in the last years of pre-university education and in practically all branches of Engineering; it is of vital importance in Design, Mechanical and Civil Engineering [1].

The main purpose of this subject is not only to provide students with theoretical knowledge of Geometry and Drawing, but also to enhance their spatial perception, one of the seven forms of intelligence and the most essential and vital one in the training of any engineer [1–3].

Although the necessity of having appropriate underpinning knowledge as well as skills and competences that can only be developed through laboratory and workshop practice is clear [4], present syllabuses allow students without this knowledge to arrive to technical degrees.

To establish the necessity of this work, a survey has been launched in a Spanish school of Engineering through a web platform used to communicate with the students to measure their previous preparation.

They were asked whether they had studied Technical Drawing at High School for two years (what is considered to be a proper preparation period of time), for only one

J. C. García-Prada and C. Castejón (Eds.): *New Trends in Educational Activity in the Field of Mechanism and Machine Theory*, MMS 64, pp. 186–194, 2019.
https://doi.org/10.1007/978-3-030-00108-7_21

year (what is considered an insufficient preparation) or if they had ever studied technical drawing before.

Surprisingly, it has been found that almost one in three Industrial Technologies students do not have the adequate preparation to face the subject with guarantees. This result is even worse in the Bachelor in Electrical Power Engineering and it becomes absolutely terrible for the Bachelor in Energy, where just four in ten students are prepared for the subject (Fig. 1).

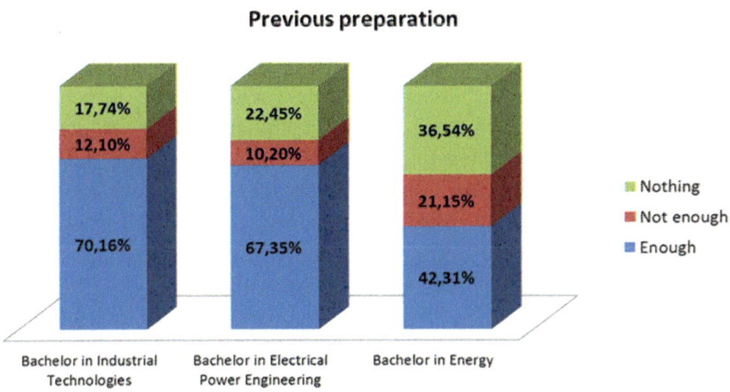

Fig. 1. Previous preparation of the Engineering Graphics students (data obtained by survey)

So the only possible solution is to help the unprepared students to reach the desired level by using e-learning. This is a solution that several teachers are currently using to solve other similar problems [5, 6].

To achieve this goal, a set of educational resources will be created. These resources will be intended for future students of the subject (trying to avoid the initial gap between them and students with a proper preparation) and to present students of the subject (trying to help them with those especially difficult contents).

Therefore, it will be necessary to find tools to develop those resources, making them attractive and fun, and easily accessible from everywhere. Also, it will be necessary to distinguish two levels, according to the two groups of students already explained.

2 E-learning

Educational innovations can improve learning outcomes and the quality of education. The effectiveness of the teaching method can be enhanced by the use of the ICTs (Information and communication technologies) [7]. From this emerges the concept of e-learning (electronic learning). The American Society for Training and Development (ASTD) defines e-learning as a broad set of applications and processes which include web-based learning, computer-based learning, virtual classrooms, and digital. Others authors delimit more the scope of e-learning, reducing only to the use of Internet [8].

But there is no doubt that it is one of the most used educational strategies at the present time not only in education, but also in professional and business context. Nevertheless, not all e-learning methodologies are completely virtual [9]. A significant proportion of university e-learning is based on blended learning (B-learning) [10–12]. In this case, virtual training is combined with face-to-face learning. Increasingly, more universities participate in such initiatives launching web portals with diversity of educational resources.

It is important not to confuse e-learning with OER (Open Educational Resources) [13]. Another thing is some online learning proposals can use different open resources. The OER are teaching, learning and research materials in any medium that reside in the public domain and have been released under an open license that permits access, use, repurposing, reuse and redistribution by others with no or limited restrictions. E-learning, as well as OER, are compatible and complementary with conventional on-site teaching and both can be used in order to improve learning aspects, such as reinforce some subjects or provide a strong academic base in some topics, inter alia.

Different types of OER or e-learning resources can be distinguished. For example, MOOCs (Massive Online Open Course) are a current educational trend in university teaching. Main characteristics of MOOC consist in they are free, massive, open, online and promote autonomous learning. MOOC also encourage the connection between users. A version of MOOC is NOOC (Nano-MOOC), ideal for segmented public who need specific training in an area. These courses have duration of up to 20 h. We may also find SPOOCs (self-Paced Open Online Course). In this case, the main difference is that the course has not time limit in order to finish it.

This paper is focused on the development of two e-learning courses in technical drawing in engineering. One of them is a payment B-learning course: a mixture between an online virtual training provided by edX platform (phase 1) combined with a face to face learning (phase 2). The other course is the result of a teacher innovation project, and would be considered as OER course, in which all the students enrolled in the subject can access for free to the whole course, but without the teacher's support.

3 Description of the Educational Resources

To meet the needs set out above, two online courses have been developed and presented in this article, covering two levels of learning: basic and intermediate.

The basic level course, referred to as "Zero Course", is designed for students who have an insufficient preparation, mostly due to not having attended any Technical Drawing course at High School. Actually, it is intended to be taught just before starting the first university course.

As regards the intermediate course, the so-called "Support Course", it is aimed at students who have difficulties in understanding specific orthographic projection topics, during the subject of Technical Drawing, taught in the first year of university.

In the following sections, the contents and methodology used in these courses are briefly presented.

3.1 Zero Course

In this course, the basics about technical drawing in engineering, emphasizing the fundamentals of the orthographic projection (or Monge's method of projection), are exposed in twelve videos in seven thematic blocks. Each thematic block consists of one or more videos and a self-evaluation test. In Fig. 2 below is the syllabus for this course (in parentheses the video number – V0 to V12 - and the duration of the video are indicated):

ZERO COURSE: TECHNICAL DRAWING IN ENGINEERING

- ❑ PRESENTATION (V0, 4'25'')
- ❑ USE OF TECHNICAL DRAWING MATERIAL (V1, 8'53'')
- ❑ REPRESENTATION SYSTEMS
  - Technical Drawing as a Language. The concept of projection (V2, 11'35'')
  - Types of Projections and Representation Systems (V3, 9'46'')
- ❑ FUNDAMENTALS OF THE ORTHOGRAPHIC PROJECTION (MONGE'S METHOD)
  - Elements of the Orthographic projection (V4, 7'51'')
  - Representation of points (V5, 15'44'')
  - Representation of straight lines and planes (V6, 8'10'')
- ❑ PARTICULAR POSITIONS OF STRAIGHT LINES AND PLANES (V7, 10'41'')
- ❑ RELATIONSHIPS OF PERTENENCE, PARALLELISM AND PERPENDICULARITY BETWEEN POINTS, STRAIGHT LINES AND PLANES (V8, 9'04'')
- ❑ INTERSECTIONS
  - straight line – straight line (V9, 2'49'')
  - Plane – Plane (V10, 4'51'')
  - straight line – Plane (V11, 5'19'')
- ❑ OBTAINING TRUE DISTANCES FROM PROJECTIONS (V12, 2'49'')

Fig. 2. Syllabus for the Zero Course (translated text from Spanish)

V1 was made with a zenith camera, so that the student can learn to use the different tools for technical drawing, and follow some examples, as if he were doing them by himself (see Fig. 3).

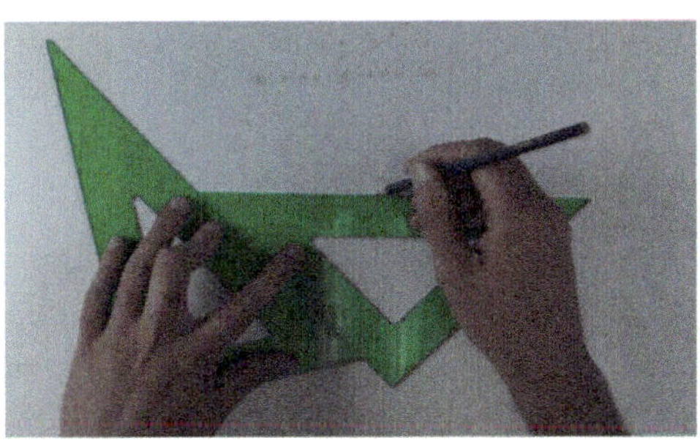

Fig. 3. Screenshot of the V1: "Use of the technical drawing material"

V2 and V3 were made with the software SmoothDraw 4 on a tablet in which one can write and draw with a special pencil as in a blackboard (see Fig. 4).

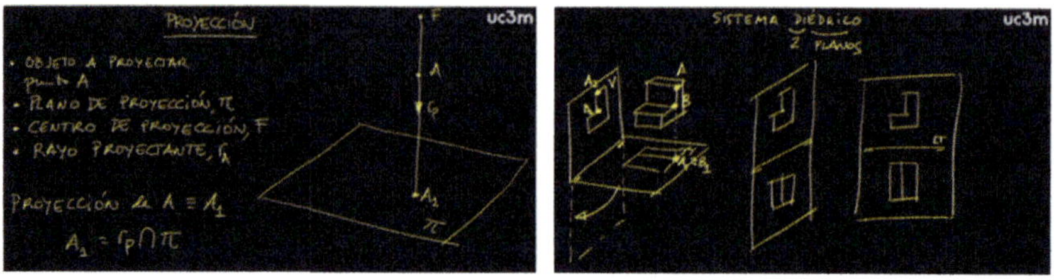

Fig. 4. Screenshot of the videos V2 and V3, belonging to the "Representation Systems" thematic block

V4 was made combining software PowerPoint and SolidEdge. The former allowed the geometrical elements to appear as the explanations were given, as well as the teacher to make annotations with the pointer; with the latter it was possible to make animations to facilitate the visualization of the projections from the object (see Fig. 5).

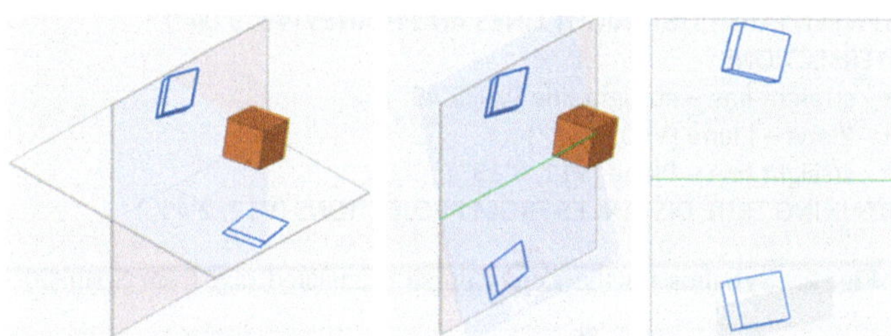

Fig. 5. Three successive screenshots from a SolidEdge animation in V4 about Orthographic Projection

The rest of the videos – V5 to V12 - were made using PowerPoint, allowing step by step explanations and annotations. Finally, all of the videos were edited with the software Camptasia.

3.2 Support Course

At this level, the educational resource has another approach: this is not a course that the student has to follow from beginning to end. It is rather a series of explanations through videos, on certain topics in which the student usually has difficulties. In this way, single videos can be viewed at any time without having seen the previous ones, even though references to other videos are indicated if necessary.

All the videos of this course were made with the software Power Point on a tablet to make annotations and drawings.

There are two types of video in the support course. The first type is composed of videos showing the main theoretical procedures of descriptive geometry, whereas the second type is made up of videos where a complex problem is proposed and solved step by step.

Both theoretical and exercise videos have three important aspects in common:

- The geometric elements and constructions involved are sequentially appearing as the theoretical topic (or complex problem) is being exposed (or solved).
- Two figures appear in the screen throughout the explanation, which makes it easier to understand: an axonometric perspective and the corresponding orthographic projections (see Fig. 6 left).

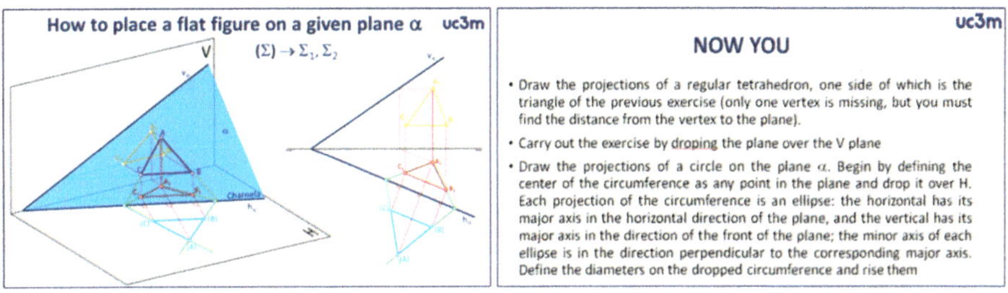

Fig. 6. Two screenshots of the video about "rotations of planes", left: Theoretical explanation with an axonometric view and the corresponding orthographic projections; right: "Now you" section. (translated text from Spanish)

- At the end of the video, in the section "Now you", the student is encouraged to deepen in aspects related to the theoretical topic just learned; or he is proposed to solve a problem based in the previous one, but of a higher level of complexity (the guidelines are given, see Fig. 6 right).

4 Students Self Evaluation Process (Zero Course Only)

A self-evaluation process has been implemented in order to measure the student knowledge acquired during the course. Main advantages of this system are:

- students with different initial preparation can learn at different paces
- it's less distracting than a group-study learning
- makes the learning deeper because the student needs to face the problem alone and he does not have the opportunity to quit thinking that the problem will be resolved by the group

A five exercises collection is associated to each theoretical chapter. These five exercises have been carefully prepared to assure that the student fully understand the subject. Once it's completed and passed the exercises collection the chapter is closed and the student is fully qualified on that matter.

Test exercises have some of the following structure:

- Objective test that have clear right or wrong answers. Multiple-choice tests fall into this group. Students have to select a pre-determined correct answer from three or four possibilities.
- Short questions that require the student to provide a definition, or a concept identification.

5 Results Analysis

At the moment, only feedback from the zero course students is available, since the Support course will be tested during this academic year that has just begun.

According to an anonymous survey, the contents of the course have been well selected. An 85% of the students, think that the level is accurate. A 12% of the think that the level is excessive (Fig. 7, left).

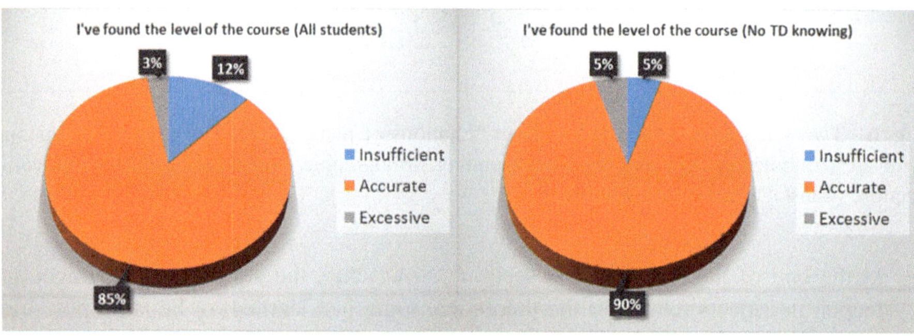

Fig. 7. Answers of the students to the question "I've found the level of the course…"

To analyze this results properly, it is necessary to know that some students who have technical drawing knowledge have applied to the course, even when it is designed for students with no technical drawing knowledge.

Separating students with technical drawing knowledge and without it, it can be concluded that 90% of the students with no knowledge find the level accurate (Fig. 7, right).

Paying attention to the created on-line material, 76% of the students think that it is also accurate (Fig. 8, left).

Restricting the response to students without technical drawing knowledge, only 67% of them find the material accurate. Almost one in four students with no knowledge, would like to have more on-line material, while 9% of them find the material excessive (Fig. 8, right). According to the opinions written by them in the survey, they have not enough time to see it all. In the other hand, 92% of the students with technical drawing knowledge find the material accurate, and no one of them thinks that it is excessive.

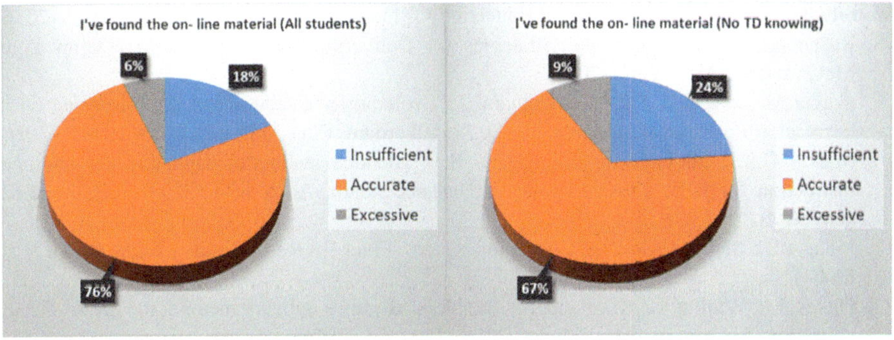

Fig. 8. Answers of the students to the question "I've found the on-line material…"

6 Conclusion and Future Work

According to students opinions, the created material is very useful for them to learn technical drawing concepts.

They think that the level of the material is accurate, but also that the amount of the subject covered by the on-line material must be extended.

Therefore, it can be concluded that the result of the work is satisfactory but there is still a lot to do in order to cover more concepts.

It is easy to propose the first future work: to extend the concepts covered by the on-line material.

Other possibility is to translate the material into English for students of the bilinguals groups.

It is also important to analyze the results of the students who have follow the zero course after their first university year. It would be very interesting to see if they have better academic results than those students that have not followed the course.

References

1. Rubio García, R., et al.: Interactive multimedia animation with macromedia flash in descriptive geometry teaching. Comput. Educ. **49**(3), 615–639 (2007)
2. Contero, M., et al.: Learning support tools for developing spatial abilities in engineering design. Int. J. Eng. Educ. **22**(3), 470–477 (2006)
3. Kaufmann, H., Schmalstieg, D., Wagner, M.: Construct3D: a virtual reality application for mathematics and geometry education. Educ. Inf. Technol. **5**(4), 263–276 (2000)
4. Tornincasa, S., Chirone, E.: The Leonardo WEBD project: an example of the WEB3D technology applications for distance training and learning. In: XIV Congreso Nacional de Ingeniería Gráfica, Santander, Spain (2002)
5. Jones, B.W.: Teaching computer graphics in the online environment. In: 59th Annual Midyear Conference of the Engineering Design Graphics Division of the American Society for Engineering Education, Williamsburg, USA (2004)
6. Rojas Sola, J.I.: Internet: La nueva puerta del siglo XXI aplicada a la enseñanza de la geometría descriptiva. Revista Mágina **10**, 197–203 (2002)

7. OECD: Innovating Education and Educating for Innovation. OECD Publishing, Paris (2016)
8. Rosenberg, M.J.: E-Learning: Strategies for Delivering Knowledge in the Digital Age. McGraw-Hill, New York (2001)
9. Garcia Marcos, C.J., Cabero Almenara, J.: Evolución y estado actual del e-learning en la Formación Profesional española. RIED. Rev. Iberoam. Educ. a Distancia **19**(2), 167 (2016)
10. Marsh, G.E.I., McFadden, A.C., Price, B.J.: Blended instruction: adapting conventional instruction for large classes. Online J. Distance Learn. Adm. **6**(4) (2003). https://eric.ed.gov/?id=EJ1068350
11. Pina, A.B.: Blended learning. Conceptos básicos. Píxel-Bit Rev. Medios y Educ. **0**(23), 7–20 (2004)
12. Dukes, M., Waring, L., Koorland, S.: The blended course delivery method: the not-so-distant education. J. Comput. Teach. Educ. **22**(4), 153–158 (2006)
13. UNESCO: Guidelines for open educational resources (OER) in higher education; 2011; 2015. (2015)

Internship Experience for Learning the Operation of a Cable-Driven Robot for Rehabilitation Tasks

G. Carbone[1(✉)], D. Cafolla[1], M. Ceccarelli[1], O. Aydinoglu[1,2], and M. Demirel[1,2]

[1] University of Cassino and South Latium, via G. Di Biasio 43, 03043 Cassino, Italy
{carbone, cafolla, ceccarelli}@unicas.it,
[2] Izmir Institute of Technology, 35430 Urla, İzmir, Turkey
{ondercanaydinoglu, muratdemirel}@std.iyte.edu.tr

Abstract. This paper describes an example of internship experience at LARM, Laboratory of Robotics and Mechatronics, at University of Cassino and South Latium. In particular, the main focus of this paper is the learning process as basis of a proper internship. Namely, a specific experiential learning approach is proposed as referring to a low-cost servo motor operation task. The learning task is further defined and developed as referring to a real application with a cable-driven robot for rehabilitation tasks, which has been designed and built at LARM in Cassino and under further investigation within Agewell project.

Keywords: Education · Lab. training · Experimental robotics Service robotics

1 Introduction

Academic internships are part of the field of experiential education aiming to apply classroom learning, theories, and experiences to professional settings. They are usually linked to an undergraduate curriculum and include reading reference sources, writing a report, critical thinking, and, especially, a real world problem solving while applying concepts from the classroom, [1].

LARM, Laboratory of Robotics and Mechatronics is an internationally recognized research center having international collaborations and exchange programs with several countries worldwide, [2]. Several academic internships have been successfully carried out throughout the last twenty years. This paper aims to describe the learning process as basis of a proper academic internship by referring to the specific experience of the first author in his internship period at LARM during the second semester of the academic year 2016–2017. The internship has been following the attendance of classrooms delivered by Prof. Ceccarelli and Prof. Carbone, which provided the necessary background in mechanisms theory and robotics, respectively.

J. C. García-Prada and C. Castejón (Eds.): *New Trends in Educational Activity in the Field of Mechanism and Machine Theory*, MMS 64, pp. 195–207, 2019.
https://doi.org/10.1007/978-3-030-00108-7_22

The learning process is further detailed in the following sections to show how a real world complex task for the operation of a cable driven robot for rehabilitation tasks has been achieved within a short eight weeks internship period.

2 Internship Arrangements

The internship has been developed in several phases as described in the scheme of Fig. 1. The first phase has been the definition of specific problem to be addressed. Namely, it has been decided the main task as being the control and operation of a set of low-cost servomotors. Then, as second phase some research has been carried out to identify the know-how and theoretical aspects of the proposed task. As third phase, specific low-cost hardware has been identified. Namely, an Arduino board has been selected, since it is very versatile and requires basic level programming and electronics skills, [3]. Similarly, specific servo-motors and electronics have been identified. Fourth phase has been learning by practicing and interactions with staff at LARM on how to integrate the selected hardware components to achieve a proper operation and control of the servo-motors. Finally, the experience learned with the above-mentioned components has been applied to CALOWI 2 (Cassino Low-Cost Wire Driven Robot version 2). This is a cable-driven robot for rehabilitation tasks, which has been designed and built at LARM in Cassino and it is under further development within the Agewell project in collaboration with Technical University of Cluji-Napoca, Romania. The specific used version of CALOWI 2 has three active degrees of freedom to be carefully synchronized to achieve a proper robot operation, [4].

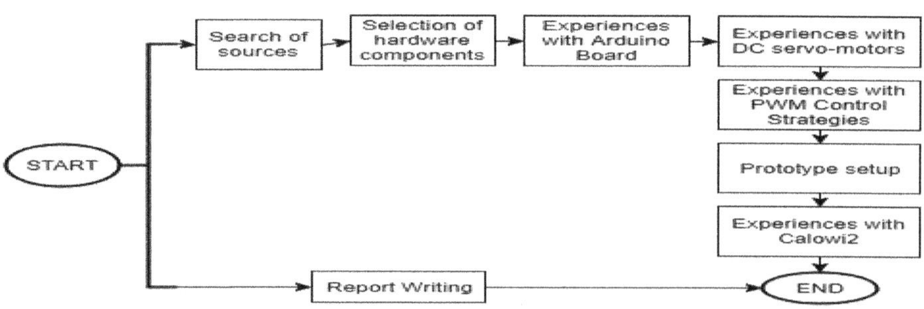

Fig. 1. A flow-chart of the learning process.

3 Arduino Control Board

3.1 Main Features

One of the main focus of the internship has been getting experienced with Arduino board, which has been chosen as the core control unit of the used prototype for this educational purpose. Arduino is a microcontroller which can be used standalone or in combination with other hardware to achieve the control of some device acquiring inputs, elaborate them, take decisions due to condition in the code and, if necessary give some output as feedback for the control loop. In this paper, an Arduino Uno has been selected and used. Arduino Uno is based on the ATmega328, [5], it has 14 digital input/output pins (of which 6 can be used as PWM outputs), 6 analog inputs, a 16 MHz ceramic resonator, a USB connection, a power jack, an ICSP header and other features are shown in Fig. 2, [6].

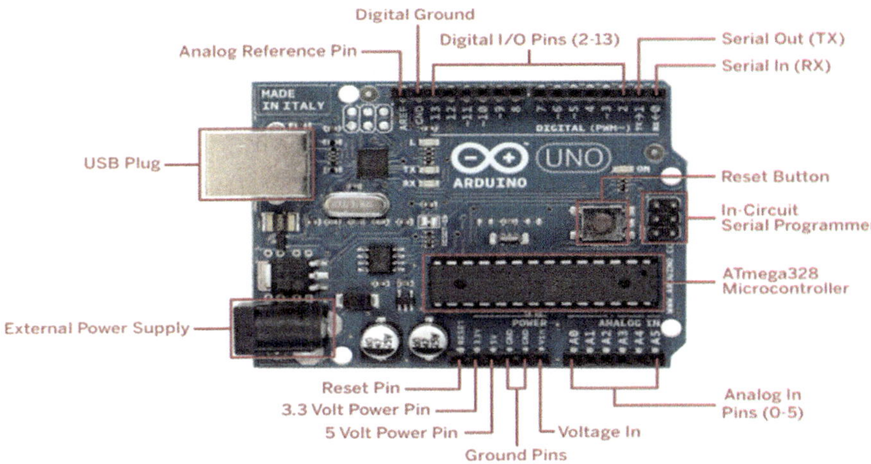

Fig. 2. A scheme of Arduino Uno, [5].

3.2 Serial Communications

Arduino can be interfaced with a PC through an Usb port. Arduino can either send commands or receive data via serial port. According to this, Arduino can receive sensors outputs and send values of variables, control robots movements and debugging program codes. Serial communication works on 1 s and 0 s (bits). These highs (1) and lows (0) bits form together and turn into bytes, after that they can be converted to ASCII encoded symbols and letters, [7].

Figure 3 shows a flow-chart for serial communication in Arduino environment.

At first Arduino's baud rate is set using the Serial.begin() function. Serial.available () checks if there is any data in the Serial port buffer. If so, Serial.read() function is used to read the data in the serial buffer. Finally, Serial.print() and Serial.println() functions visualize in the Serial monitor interface whatever has been stored in the buffer. These functions can help the debug procedure to check visually inputs and outputs, and communication between the control unit and the user, [8].

3.3 Data Parsing

Data parsing is useful when sending several information at the same time on a buffer is needed, for example a vector of position that an actuator has to reach such as in Fig. 4. The parsing procedure will analyze this flux of data and assign it to the needed

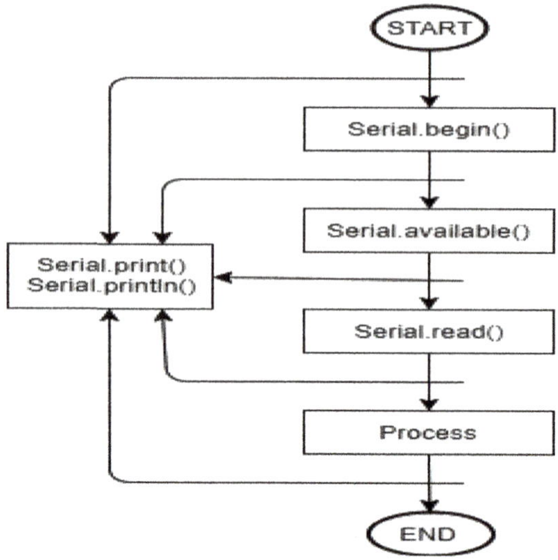

Fig. 3. An example flow-chart for serial communication.

variables. For this project, the parsing method has been chosen synchronizing the motion of different servo-motors. The positions are sent in a data flux, divided and converted to integers using .toInt() command, [9]. An example of data parsing vector position is shown in Fig. 4 where 30,60,415 instruction stands for a target position of 30°, 60°, 415° for motor 1, motor 2 and motor 3, respectively.

Fig. 4. An example of data parsing.

4 Using Servo-Motors

A further step during the internship has been the implementation of a control strategy to move the chosen servo-motors to achieve the experiments target.

A servo-motor is a DC motor which is controlled for specific angular rotation with the help of additional servomechanism by means of a closed loop feedback control system. Servo motor applications are seen in remote controlled servomechanisms for controlling the direction of motion. Furthermore, they can provide angular precision, [10]. A servo-motor presents three cables namely, brown, red and yellow. The brown cable must be connected with the common ground (GND), the red one must be connected with the power supply (5 V) and the yellow one must be connected with the control signal, which is provided through a pulse width modulation (PWM), [5].

4.1 Pulse Width Modulation (PWM) Strategy

Servo control is achieved by sending to a servo a pulse-width modulation (PWM) signal. Namely, a series of repeating pulses of variable width where either the width of the pulse or the duty cycle of a pulse train determines the position to be achieved.

When a square wave is sent, 5 V is applied in the "on" position; 0 V is applied in the "off" position. The width at which the "on" part is active is called "Pulse Width", [11]. The parameters to set the pulses are the minimal pulse width, the maximal pulse width, and the repetition rate or duty cycle. Given the rotation constraints of the servo, neutral is defined to be the position where the servo has exactly the same amount of potential mechanical rotation in the clockwise direction as it does in the counter-clockwise direction. Usually, the neutral position is related to 1.5 ms pulse width. A graphical example of PWM is shown in Fig. 5.

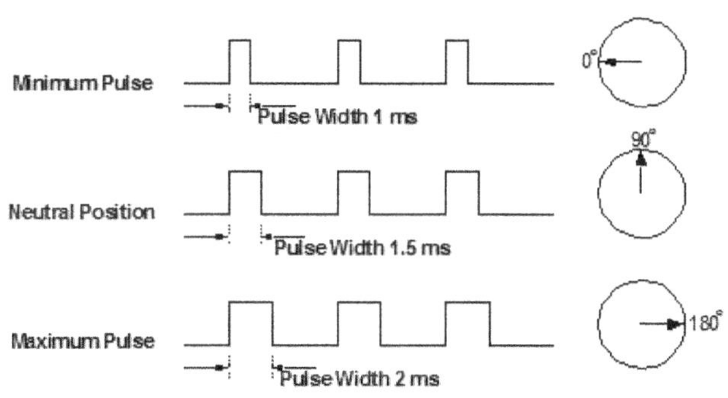

Fig. 5. An example of PWM signals with different duty cycles.

4.2 Testing Servo-Motor Rotation

Experimental activity has been carried out to operate three HS-785HB servo motors by using an Arduino Uno board. As a first step both motors data sheets and real PWM scales are checked with the help of an Arduino code. This Arduino code has been set up by using the commands in Sect. 3 (Sending data, PWM, Serial Monitor). It is to note that servo-motors usually have an angular motion range of ±180°. HS-785HB servo motors have an angular motion range of ±1332°. Max PWM Signal Range, Max Travel Angle and No-Load Speed at 4.8 V are taken from datasheet [12] and compared with real values as reported in Table 1.

Table 1. Comparing Datasheet and Real values.

	Datasheet	Real
Max PWM Signal Range	600–2400 μs	615–2415 μs
No-Load Speed (4.8 V)	1.68 s/360°	1.57 s/360°
Max Travel Angle	2826°	2664°

4.3 Adafruit 16-Channel Servo Driver

Microcontrollers such as Arduino board usually have a limited number of PWM outputs and a limited computational power. If a prototype presents too many actuators, signal problems can occur. To overcome these problems, an Adafruit 16-Channel 12-Bit PWM /Servo Driver has been selected and used in this project in cooperation with an Arduino Uno board. With this hardware it is possible to obtain up to 16 PWM outputs using only 2 pins on the Arduino controller. This hardware solution keeps other pins available for scaling up the hardware, for example, to increase the number of controlled servo-motors of for the operation of other devices. Furthermore, Adafruit 16-Channel 12-Bit PWM /Servo Driver uses an I2C protocol, which allows a common clock and synchronous operation of all the connected servos. Accordingly, Adafruit_PWMServoDriver.h and Wire.h libraries have been used instead of Servo.h Arduino library, for an optimal communication between the servo driver. Moreover, the function pwm.setPWM(pin number, ON, PWM) has been used to send the position to the actuators, [13].

Figure 6 shows an example of how Arduino, a servo-motor, the Adafruit servo driver and power supply are connected. It is important to note that Adafruit 16-Channel 12-Bit PWM /Servo Driver also simplifies the cabling of servo-motors, since a single power supply unit is needed for all the connected servos.

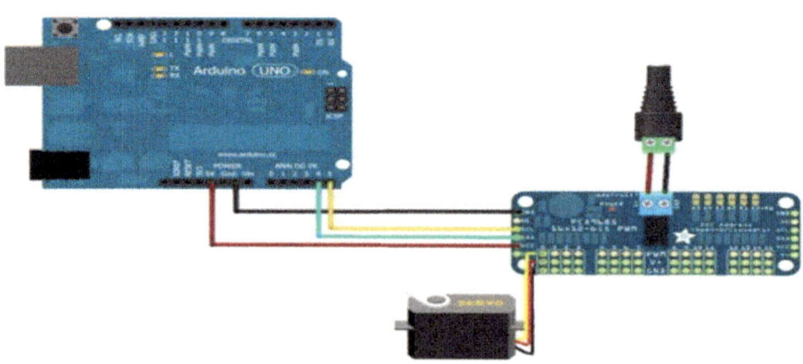

Fig. 6. Adafruit Servo Driver servo-motor connection example, [13].

5 Synchronized Control of the Servo-Motors

Robot manipulators are composed by links and joints. Links are the rigid members connecting joints or axes. The axes are the movable components of the robotic manipulator that cause relative motion between interconnected links, [14]. Manipulators must have synchronized motion of each joint to achieve a proper target operation task.

Experimental tests have been carried out aiming to achieve a synchronized control of three servo-motors with different angular configuration targets. At first a calibration procedure has been carried out to map the servo-motors in Arduino environment. Namely, servo-motors max and min values have been set as 158 and 660 PWM, respectively. Moreover, 0.18 ms PWM pulse width has been set as equals to 1° of rotation. Starting position has been set in the middle of the servo-motor motion range. 0° corresponds to a PWM of 400 and the range is between +1332° and −1332°.

The developed program code includes three different angles that are defined as servo1angle, servo2angle and servo3angle.These angles are converted to PWM values, which are defined as servo1pwm, servo2pwm and servo3pwm by using Eqs. (1–3).

$$\text{servo1pwm} = (\text{servo1angle} * 18)/100 + 400 \tag{1}$$

$$\text{servo2pwm} = (\text{servo2angle} * 18)/100 + 400 \tag{2}$$

$$\text{servo3pwm} = (\text{servo3angle} * 18)/100 + 400 \tag{3}$$

There are two possible cases for servo1pwm. It can be servo1pwm > 400 and servo1pwm < 400. This means that user command to servo to rotate clockwise or counterclockwise. For identifying these two cases, the developed program code includes two if loops. In each if loop there is a for loop for step counting. Each for loop determines how many steps are needed to complete the desired motion with constant PWM. This constant is 1° = 0.18 PWM to achieve a smooth motion. In the same for loop number of steps for the second and third servo-motors will be set as the same of servomotor1 but angular step rotation will be different according to factor 2 and factor 3, respectively. Equations (4–5) can be used for finding the step rotation of second and third motors.

$$\text{factor2} = (\text{servo2pwm} - 400)/((\text{servo1pwm} - 400)/0.18) \tag{4}$$

$$\text{factor3} = (\text{servo3pwm} - 400)/((\text{servo1pwm} - 400)/0.18) \tag{5}$$

The programming process is also described in the flow-chart shown in Fig. 7 where three servo-motors can reach different target angular positions at same time with same step number. In addition, in this code data parsing method has been used to receive a vector of servo rotation angles from the serial monitor interface.

In one experiment angles were defined as servo1angle = 90°, servo2angle = 180° and servo3angle = −90°. In their direction servo 1 rotates about 90° clockwise, servo 2 rotates about 180° clockwise, servo 3 rotates about 90° counter clockwise. Figure 8 shows the synchronous motion experiment target positions.

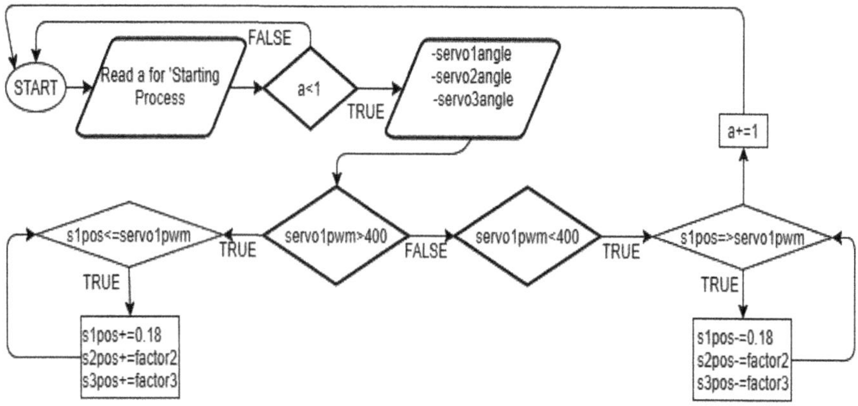

Fig. 7. A flow-chart of the operation of three servo motors.

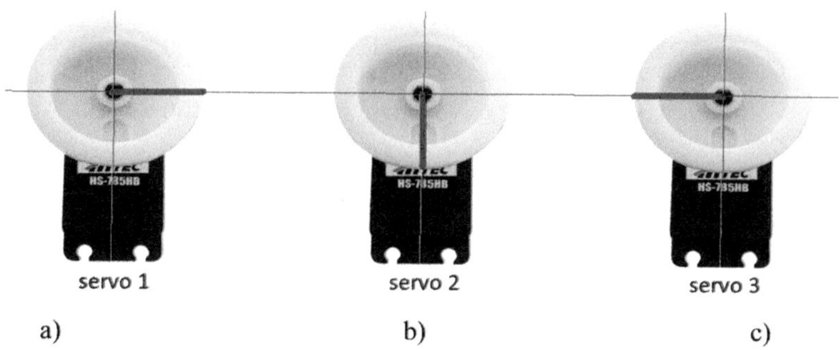

Fig. 8. Three servo-motors with position marks: (a) servo 1 at 90° clockwise position; (b) servo 2 at 180 clockwise position; (c) servo 3 at −90° counter clockwise position.

6 A Case of Study with CALOWI 2

CALOWI 2 robot is a cable-based parallel robot for rehabilitation, which has been designed and built at LARM in Cassino, [4]. The structure of CALOWI 2 is a non-conventional open architecture, which allows an easy accessibility by patients under treatment. Aluminum profiles are used so that the robot is light and stiff while it can be assembled and disassembled easily for storage and transportation. Cables are connected to the end-effector which covers the arm to be trained by using a wristband, [4]. During the rehabilitation, an arm needs elbow support. On CALOWI 2 robot has three servo motors and three cables, which are connected to manipulators and the end-effector through a wristband as shown in Fig. 9. Cable tensioning is guaranteed by gravity. The success of a rehabilitation process requires several key factors. Among them, it is necessary to identify and perform a reliable and safe motion training protocol. For the purpose, several experiments have been carried out to set up proper motion parameters including but not limited to the elbow support position and arm motion ranges. Some tested setting conditions are shown in Fig. 10.

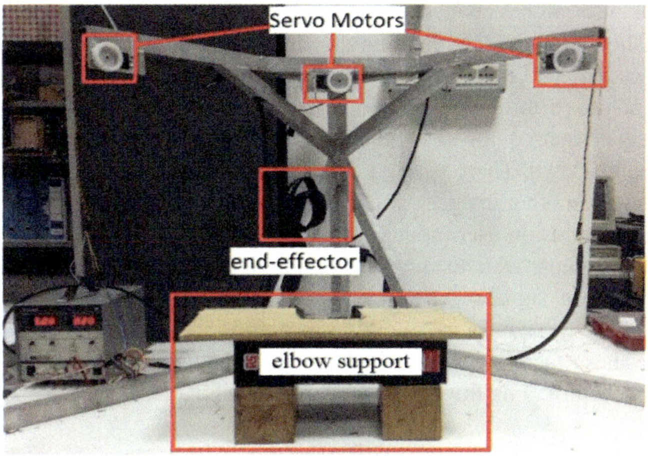

Fig. 9. A picture of CALOWI 2 cable-driven parallel robot at LARM.

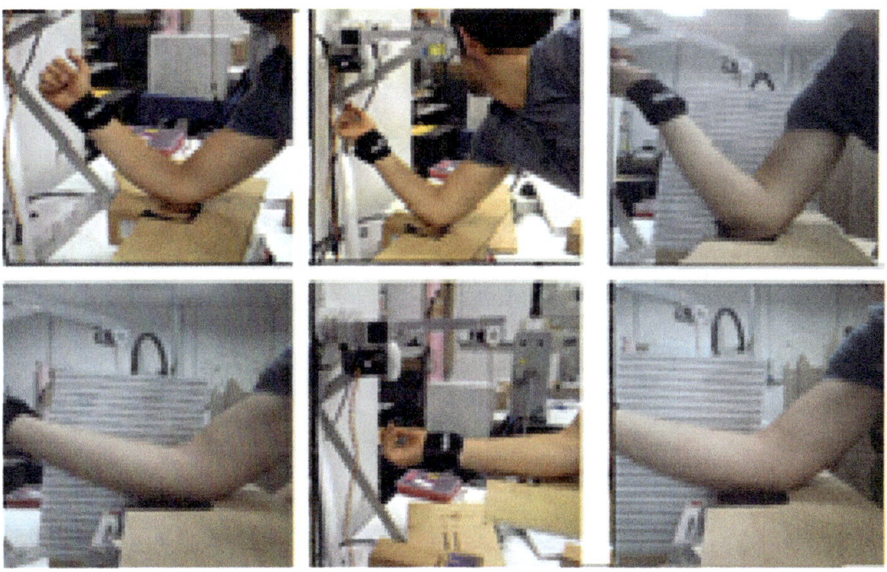

Fig. 10. Several position testing for identifying most favorable arm supporting configuration.

6.1 Programming CALOWI 2 Operation

Sections 3 and 4 have defined the main hardware and software features for the operation of the three DC servomotors, which are used to operate CALOWI 2. As main operation tasks it has been decided to achieve two operation tasks consisting in assisting the motion of a human arm during up-down and left-right motions. Motion distances are constant for each motion.

A specific user interface has been designed on a laptop screen and keyboard interface. At beginning the screen shows a message ‘Which motion?’ followed by ‘How many cycle?’ and then a ‘Starting process’ button becomes available. The first command is achieved through a for loop. There is an integer which is defined as ‘a’ variable. If this integer equals to 1, the motion will be left-right, else if equals to 2 the motion will be up-down. There is also an integer variable called as ‘c’. If c equals 1 the process will start. Another integer variable ‘b’ is defined to determine the cycle number and this value depends on user input with three possible cases (c = 1; c <= b; c += 1). The three servo-motors start to move in middle angle position (s1pos = s2pos = s3-pos = 400 ms PWM signal). After that, s1pos increases or decreases with constant value (0.18 ms PWM signal), meanwhile s2pos and s3pos increases or decreases with their factors. These factors (factor2, factor3) are calculated at beginning of the algorithm. A detailed programs flow-chart is shown in Fig. 11.

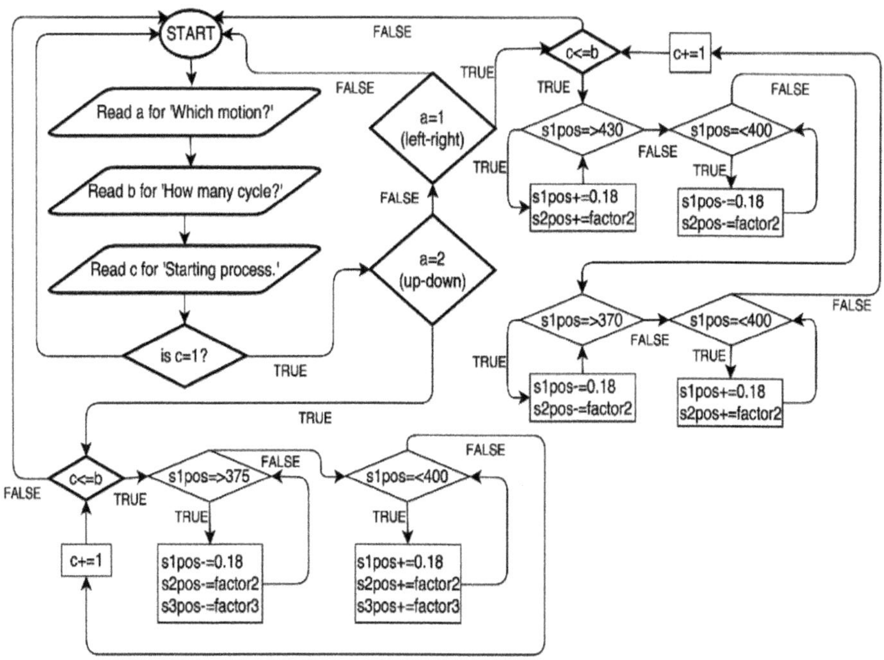

Fig. 11. A CALOWI 2 operations flow-chart.

6.1.1 Left and Right Motion

One cycle of left and right motion takes four second. The motions start position is at middle configuration. End-effector goes to the left position (10 cm) in one second and turns back to start position finally end-effector goes to the right position (10 cm) and turn back to start position. Experimental tests have been first carried out without connecting CALOWI 2 to a human in order to verify a suitable safe operation is successfully shown in Fig. 12. Then, experiments have been taken by assisting the motion of a human arm as successfully shown in Fig. 13.

a) b) c)

Fig. 12. Experiment of left and right motion without hand: (a) t = 1 s left side; (b) t = 0 s initial position; (c) t = 3 s right side.

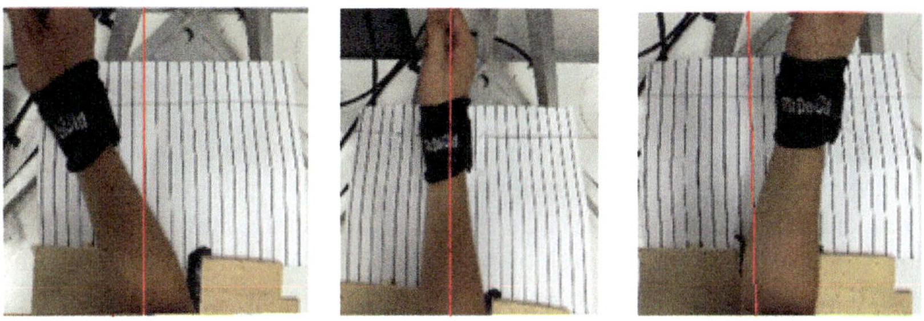

a) b) c)

Fig. 13. Experiment of left and right motion with hand: (a) t = 1 s left side; (b) t = 0 s initial position; (c) t = 3 s right side.

6.1.2 Up and Down Motion

One cycle of down and up motion takes 90 ms. The motions start position is on middle. End-effector goes to the up position (15 cm) in 75 ms and turns back to the start position by taking the same amount of time. Experimental tests have been first carried out without connecting CALOWI 2 to a human in order to verify a suitable safe operation is successfully shown in Fig. 14. Then, experiments have been taken by assisting the motion of a human arm as successfully shown in Fig. 15.

a) b) c)

Fig. 14. Experiment of up and down motion: (a) t = 0 s initial position: (b) t = 0.5 s third point; (c) t = 0.75 s fourth point.

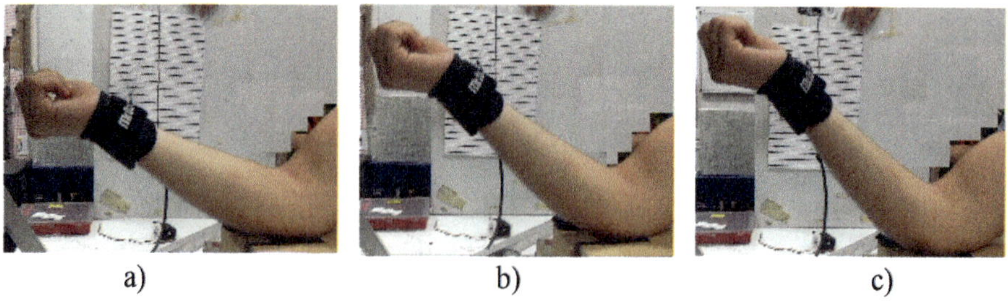

Fig. 15. Experiment of up and down motion: (a) t = 0 s initial position; (b) t = 0.5 s third point; (c) t = 0.75 s fourth point.

7 Conclusions

This paper describes an academic internship which has been successfully carried out at LARM. Main aim of this paper is to show the learning by experience process, which is achieved during an internship. The reported case of study shows how it has been possible to achieve the complex task of operating CALOWI 2, a cable driven robot for rehabilitation tasks, within a short eight weeks internship period.

Acknowledgements. The paper presents results from the research activities of the project ID 37_215, MySMIS code 103415 "Innovative approaches regarding the rehabilitation and assistive robotics for healthy ageing" co-financed by the European Regional Development Fund through the Competitiveness Operational Programme 2014–2020, Priority Axis 1, Action 1.1.4, through the financing contract 20/01.09.2016, between the Technical University of Cluj-Napoca and ANCSI as Intermediary Organism in the name and for the Ministry of European Funds.

References

1. Merritt, R.D.: Student internships. EBSCO Research Starters: Academic Topic Overviews, pp. 1–8 (2008)
2. Ceccarelli, M.: An Illustrated history of LARM in Cassino. In: International Workshop on Robotics in Ale-Adria-Danube Region RAAD 2012, Napoli, pp. 35–42 (2012)
3. Arduino Homepage: Arduino guide. https://www.arduino.cc. Accessed 8 Sept 2017
4. Carbone, G., Gherman, B., Ulinici, I., Vaida, C., Pisla, D.: Design issues for an inherently safe robotic rehabilitation device. In: 26th International Conference on Robotics in Alpe-Adria-Danube Region RAAD 2017, pp. 967–974. Springer, Dordrecht (2017)
5. Banzi, M., Shiloh, M.: Getting Started with Arduino: The Open Source Electronics Prototyping Platform. Maker Media Inc., Sebastopol (2014)
6. Balogh, R.: Educational robotic platform based on Arduino. In: 1st International Conference on Robotics in Education RiE2010, FEI STU, pp. 119–122 (2010)
7. Hisham, A.A.B., Ishak, M.H.I.: Bluetooth based home automation system using an Android phone. Jurnal Teknologi (Sci. Eng.) **70**(3), 57–61 (2014)
8. Margolis, M.: Arduino Cookbook: Recipes to Begin, Expand, and Enhance Your Projects. O'Reilly Media Inc., Beijing (2011)

9. Thu, S., Tin, H.H.K.: Construction of Android and Arduino for home appliances switching system. Int. J. Eng. Appl. Sci. Technol. **1**(11), 66–71 (2016)
10. Sawicz, D.: Hobby servo fundamentals. Princeton ebook. https://www.princeton.edu/~mae412/TEXT/NTRAK2002/292-302.pdf. Accessed 8 Sept 2017
11. Sparkfun Homepage. https://learn.sparkfun/PWM. Accessed 8 Sept 2017
12. Servocity Homepage. https://www.servocity.com/hs-765hb-servo. Accessed 8 Sept 2017
13. Adafruit Homepage. https://cdn-learn.adafruit.com. Accessed 8 Sept 2017
14. Ceccarelli, M.: Fundamentals of Mechanics of Robotic Manipulation. Springer, Dordrecht (2004)

Free and Open Source Software for Technical Texts Editing, Its Advantages and Experience of Usage on TMM Training in Bauman University

Alexander Titov(✉) and Andrei Vukolov

Aerospace Faculty, Bauman Moscow State Technical University, Moscow, Russian Federation
aserwqq@gmail.com, twdragon@bmstu.ru

Abstract. Creation of the documentation is practically hardest part of any engineering, software development or scientific project containing large amount of knowledge. In Bauman University every TMM project includes main document (thesis) which contains explanation of all calculations performing through development of the project. To reduce time of preparation for such technical texts the complete software solution with formulae support needed. In this paper usage of LaTeX (as common free solution) is considered in comparison with most known free and proprietary solutions such as MS Word, Open Office and T-Flex DOCs.

Keywords: Open source · Free software · TMM
Engineering education · Text documentation · Technical texts
LaTeX · MS Word · Libre Office · Open Office · CAD/CAM

1 Introduction

Technical progress made computer text editing solutions the only useful instruments for technical text preparation in engineering education. However, most solutions in this area are proprietary. Such products are very difficult to use in permanently changing environment, for example in shared computer classes. They require licensing control services that are often difficult to support through many machines. Also when educational organization owns more than 3–4 different proprietary solutions it is always big problem to keep classes ready. It happens because license control solutions from different vendors consume large amount of computational resources and often require newest versions of operating systems. The special problem is incorrect inheritance of license requirements. It can make it impossible to deploy documentation and/or instruments developed by student onto his own machine or the machine placed outside main university's network.

J. C. García-Prada and C. Castejón (Eds.): *New Trends in Educational Activity in the Field of Mechanism and Machine Theory*, MMS 64, pp. 208–215, 2019.
https://doi.org/10.1007/978-3-030-00108-7_23

To avoid described problems and constraints many open-source communities such as GNU project developed special documents called *Free Licenses* [1,2,5,6]. These documents allows user who use the software issued under free licenses, to develop and distribute his own programs, documentation and solutions based on, and place texts, code and/or executables into public domain with well-known requirements easy for understanding [19]. Clearing and granting function for permission of usage of free licenses is carried by Free Software Foundation (FSF) and Open Source Initiative (OSI) organizations. Now usage of free software grows active through all over the world [14] including Bauman University [9,23]. Besides free licenses the engineers can distribute products, graphics and documentation freely without permission of redistribution.

The most common types of texts developed in Bauman University are specification, development thesis and datasheet. Preparation of each type of document is an important part of educational process. In typical environment it requires several specialized software solutions. In Bauman University the specifications are usually prepared using the same software than the graphical documentation [9,23,24]. Development thesis and datasheets are usually prepared using common office solutions. On this stage, however, it is important to disconnect the process with concrete solution because the student goes to achieve possibility to *develop a system*, not to *use a program*.

2 Peculiarities of Text Documentation Developed in Bauman University

Text documents prepared by students in Bauman University contain bibliography, large amount of graphics and formulae (Fig. 1). For ordinary project there are no special requirements on software. But the main solution which are in active usage on almost all faculties and departments is *Microsoft Word*. It is proprietary solution which can not be used outside university's network. Some documents also require special fonts and typesetting such as ISO or GOST font kits. These typesetting kits are also proprietary and provided by such vendors as *Autodesk* and *TopSystems*. The technical texts in Bauman University often

Fig. 1. Typical development thesis pages

required to be prepared in PDF format. Windows environment does not provide any native solution for conversion from *Word* native DOC format to PDF instead of proprietary *Adobe Acrobat.* This possibility is provided in open source office suite named *LibreOffice* with some restrictions: for example, it is not possible to embed multimedia content into PDF file using free software. After described pipeline the developed document is strictly connected with solution: the license places restrictions on any case of its usage. Possibility of the free redistribution of developed documents can be obtained only with free pipelines.

3 Advantages of Free Software in Training Process

Here and below the advantages of free software in comparison with proprietary solutions will be given for the following list of features:

1. Cross-platform interoperability;
2. Processing of formulae;
3. Processing of raster graphics;
4. Generating and processing of vector(ized) graphics.

3.1 Cross-platform Interoperability

The first step of text document preparation is the page and exterior typesetting. WYSIWYG[1] solutions like *Word* or *LibreOffice* provide special dialogs This type of document type setting requires to repeat configuration for each new document which are being prepared using the solution. To avoid this some WYSIWYG solutions, including *Word* provide template mechanism which allows user to create several documents inherited from single special prepared one called *document template* (.DOT-file). This model makes document dependent on the solution and even on its version because the templates prepared within different versions of the same program would be uncompatible due to embedded VBA core.

Unlike this, LaTeX does not follow WYSIWYG methodology. It provides TeX programming language for type setting and markup configuration tasks. In this case the document consists of plain text only and it has special part called *preamble* which holds the same functions as the template: to describe minor changes of the exterior that aren't inherited:

```
\documentclass[graybox,stagethree,]{svmult}
\usepackage{amsmath,amssymb,url,listings}
\usepackage{epsfig,epstopdf}
\makeindex
\graphicspath{{img/}}
```

All other type setting instructions are encapsulated within the special *style file.* It is not a template but the processing rules which are strictly prescribed by LaTeX language standard. This variant is truly cross-platform because all files as

[1] **WYSIWYG** [3] ("What You See Is What You Get")—document preparation methodology which prescribes identical view of the document on editor's screen and within any other type of view.

consist of the plain text as document itself [4]. The same set of source files produce also the same result on all operating systems and hardware platforms which LaTeX processing engine had been realized on. Also LaTeX processing engine may produce output in both PostScript, PDF or DVI formats. PDF format now is truly cross-platform because it has full support from several vendors including free software projects. At the same time, *Word* and *LibreOffice* provide proprietary DOC format and free but not well known OpenDocument format respectively. These facts and especially uncompatibility of *Word* with Linux [24] make LaTeX practically the only true cross-platform solution for text documentation preparation.

3.2 Formulae Processing

Formulae typographic configuration is very hard task to be solved by ordinary office solutions which oriented on inline text editing model [17]. The WYSIWYG solutions are forced to divide the text editing and formulae processing into the independent tasks. In that case the formula is defined as *smart object* which has its own structure and processing rules. The special program producing such objects provides graphic representation of the formula as a bitmap image encapsulated (not always) with source code of produced object. This graphic representation can be interpreted as the special symbol within the text. In is not a vector image and requires to be re-rendered each time user changes its dimensions.

All known WYSIWYG office suites implements special technologies to create *links* from the text document to the formulae object code. Among the others, the office suites which use file format contains compressed binary data with hierarchic structure, like *Word* and *LibreOffice* provide most user-friendly interface. They allow to embed the formula directly into the document and to reserve required data space within the file. This technology, known as OLE[2] in Microsoft's variation [7,20,21] requires the special software called OLE servers for each type of object that can be embedded. Each one of them requires special integration into the specific solution. *Word* and *Microsoft Office* suite provides built-in OLE formulae processing server called *Microsoft Equation* (Fig. 2).

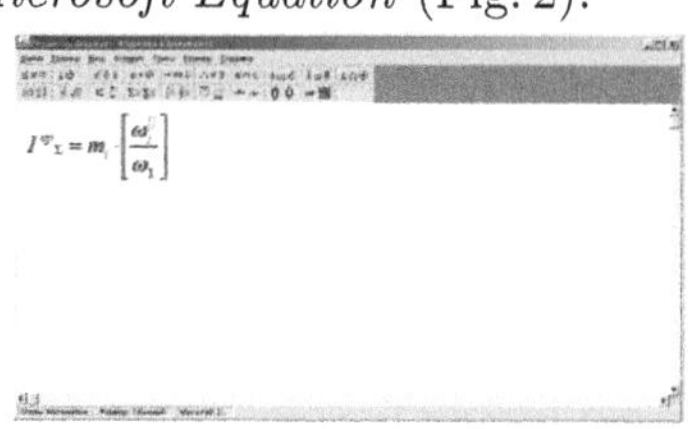

Fig. 2. *Microsoft Equation 3.0* interface

This model of formulae processing carries all disadvantages described above. Also it leads to impossibility to transfer the prepared document between even different installation of the same software because of possible difference between versions of the server (*LibreOffice* suite does not have this disadvantage because it provides encapsulated server named *LibreOffice Math*). The process of creation of a formula in visual environment which looks like shown on Fig. 2 is extremely slow because user is forced to permanently switch between keyboard

[2] Object Linking and Embedding.

input and graphically invokable commands that produce symbols and placeholders on screen. Practically the main and the only advantage of the visual formulae editing in the training process is good interaction between the object and user actions.

In LaTeX the formula is not an object [22]. It consists from usual symbols of the required fonts rendered into the geometrical structure [12]. The embedding of the formula does not required in this case. LaTeX renders it as plain text printed using special set of fonts. The formula rendered on Fig. 2 in LaTeX looks like:

$$I_\Sigma^{pr} = \left[\frac{\omega_i}{\omega_1}\right] I_i$$

and can be represented in document's source code as the following set of commands:

```
\begin{equation*}
I_\Sigma^{pr} = \left[\frac{\omega_i}{\omega_1}\right]I_i
\end{equation*}
```

The source text in the document contains formula divided into the set of mathematical objects presented as commands and expressions. The LaTeX core program performs rendering itself. Definitions inside the formula represent the consistency and sequence of calculations defined and ordered by statements of TeX language. This model of formulae processing allows to easily connect mathematical objects from text document to real calculation engine. For user the formula structure in LaTeX has very strong connection with structure of document itself. In training process it carries some very important advantages:

- Creation of the formulae in LaTeX is significantly faster than in WYSIWYG software suites because it is not necessary for user to switch between viewing of result and editing the expression.
- Sequence of expressions allows student to improve understanding of logic of calculations in TMM.
- Programmable structure of the formula allows student to easily use once prepared mathematical expression again and again, and even to build a database of formulae ready-to-use for further projects.

The only significant disadvantage for LaTeX formulae processing engine in TMM training is necessity to learn commands to build a formula. But existence of this disadvantage can easily be transformed into an advantage according to famous "Russian Method" [8,13] of engineering education. In this case the student improves his skills in programming, which also included in the course, simultaneously with performing the calculations by himself. After that he will have an additional skill which allows him to develop special project (IDE[3] or report generator for example) where the formula can be generated automatically using LaTeX commands, self-developed program and/or semantic database [15].

---

[3] Integrated Development Environment.

3.3 Graphics Creation and Processing

WYSIWYG solutions treat raster and vector graphics separately. Most common way to treat the raster images is to use embedding technology as it was described above. Rendering and unpacking of bitmaps are server's tasks in this case. For vector graphics there are special solutions integrated into the software suite [25]. These solutions perform limited operations with vectorized objects and all of them are embedding servers (Fig. 3) which produce raster graphics from vector object's source code before rendering it into the document. This method is absolutely incompatible with engineering education process because the obtained drawings can not be read using any standard CAD system or technical assistance software. Also WYSIWYG solutions contain no tools to satisfy the technical standards because it is not their primary function. The considered facts allows us to make a conclusion: preparation of technical documents containing graphics, especially vectorized, is very difficult using typical office software, even for engineers, not only for students.

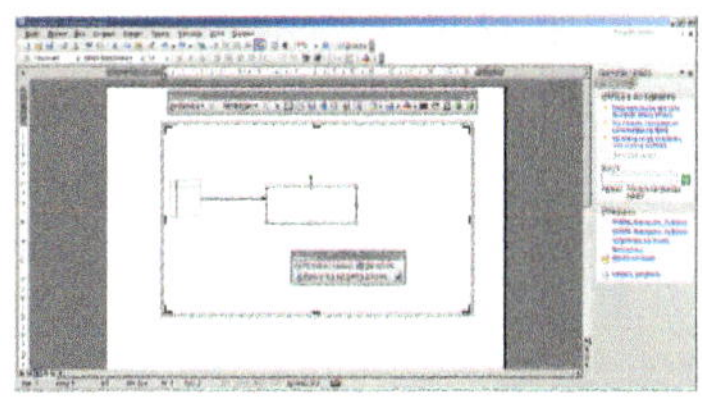

Fig. 3. *Microsoft Word 2003* OLE built-in vector drawing engine interface

Some solutions, like *T-Flex DOCs*, provide complex automation of technical texts creation and editing [10]. But for TMM training process these solutions are completely unsuitable because they all require special skills and authorized training courses for operators [16]. Within TMM course which completes in one year it is fully impossible to use such systems.

In LaTeX the graphics rendering engine realized as set of modules which can be used separately. Each module, for example, `graphicx` [11] performs rendering of specific type of graphics. The possibility of direct rendering makes LaTeX engine truly cross-platform as it was already described above. But the most significant advantage of LaTeX graphic engine is its availability to render programmable vectorized graphics using scripts. Some modules like TikZpicture and PGFplots [18] allow user to create a script which defines plotting settings, including frame, legend, curve appearance. After that it is possible to transfer data from mathematical expression or file generated by another program (e.a. in CSV format), directly to the prepared script to render the plot. The short example shown on Fig. 4 renders 3D surface using mathematical expression defined as:

```
\begin{tikzpicture}
        \begin{axis}[view={110}{10},colormap/greenyellow,colorbar]
                \addplot3[surf] {-sin(x^2 + y^2)};
        \end{axis}
\end{tikzpicture}
```

The described possibilities make LaTeX practically the complete helper solution for preparation of technical texts including sophisticated tasks of generating vector graphics, script and array-organized data based plots rendering. During TMM training student can easily satisfy requirements of Russian method using experimental data and updating it directly while document preparation. It connects real development and training allowing university to improve alumni's competences.

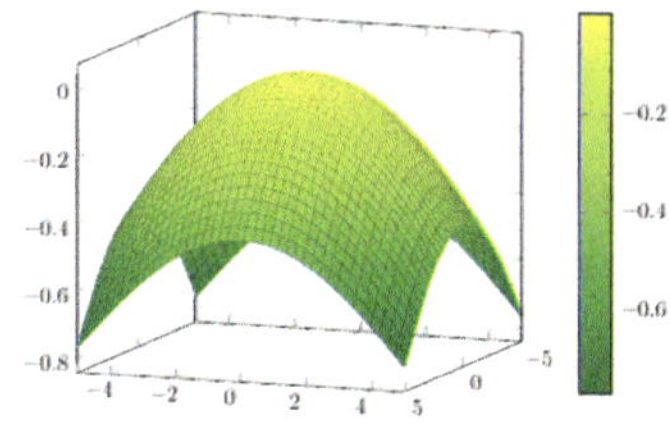

Fig. 4. PGFplots render example

4 Conclusions

Now LaTeX is in active usage in Bauman University as main free and open source text editing software. It replaces many proprietary solutions like *MS Word* and allows students to highly improve their understanding of engineering work using TeX programming, building of the calculation sequences and logical skills. The considered facts and peculiarities described above made LaTeX the ideal solution for text documentation preparation in Aerospace faculty of Bauman University.

Acknowledgements. Authors want to acknowledge prof. **Olga Egorova** for her assistance and advice during preparation of this paper.

References

1. Common development and distribution license (CDDL-1.0). Published by Open Source Initiative. https://opensource.org/licenses/CDDL-1.0
2. The MIT license. Published by Open Source Initiative. https://opensource.org/licenses/MIT
3. WYSIWYG, p. 1936. Springer, Boston (2001). https://doi.org/10.1007/1-4020-0613-6_21274
4. Erste Schritte, pp. 13–46. Springer, Heidelberg (2006). https://doi.org/10.1007/978-3-540-34584-8_2
5. GNU GPL general public license version 3, 29 June 2007. Published by Free Software Foundation (2007). http://www.gnu.org/licenses/gpl.html
6. GNU LGPL lesser general public license version 3, 29 June 2007. Published by Free Software Foundation (2007). http://www.gnu.org/copyleft/lesser.html
7. Bangsow, S.: Data exchange and interfaces, pp. 661–704. Springer, Cham (2016). https://doi.org/10.1007/978-3-319-19503-2_14
8. Baryshnikova, O.O., Egorova, O.V., Omelchenko, V.: An efficient educational methodology for teaching theory of machines and mechanisms discipline. In: Proceedings of 14th IFToMM World Congress 2015, Taipei, Taiwan, vol. 1 (2015). https://doi.org/10.6567/IFToMM.14TH.WC.OS5.008, http://www.iftomm2015.tw/IFToMM2015CD/PDF/OS5-008.pdf

9. Belonozhko, P.P., Belous, V.V., Karpenko, A.P., Khramov, D.A.: Instruments for automated numeric estimation of trainees' meta competences. Sci. Educ. Electron. Sci. J. (10) (2015). (in Russian). http://technomag.bmstu.ru/doc/821623.html
10. Bocharnikova, N.A.: Complex automation of Uralkryomash AO using PLM-system. Autom. Ind. **9**, 7–10 (2016). (in Russian)
11. Carlisle, D.P.: The LaTeX 3 project. Packages in the 'graphics' bundle. http://mirrors.ctan.org/macros/latex/required/graphics/grfguide.pdf
12. Cohl, H.S., Schubotz, M., McClain, M.A., Saunders, B.V., Zou, C.Y., Mohammed, A.S., Danoff, A.A.: Growing the digital repository of mathematical formulae with generic LaTex sources, pp. 280–287. Springer, Cham (2015). (in Russian)https://doi.org/10.1007/978-3-319-20615-8_18
13. Egorova, O.V.: About "Russian Method" of training of engineers. Int. Aff. (1) (2014). (in Russian). http://interaffairs.ru/read.php?item=10459
14. Ghosh, R., Glott, R., Krieger, B., Robles, G.: Free/libre and open source software: survey and study. Technical report, International Institute of Infonomics, University of Maastricht (2002)
15. Groza, T., Handschuh, S., Möller, K., Decker, S.: SALT—semantically annotated LaTeX for scientific publications, pp. 518–532. Springer, Heidelberg (2007). https://doi.org/10.1007/978-3-540-72667-8_37
16. Kochan, I.N.: Complex automation based on T-FLEX PLM platform. CAD Graph. **6**(224), 54–56 (2015). (in Russian)
17. Lin, Y., Zini, E.: An empirical study on implementing free/libre open source software (FLOSS) in schools, pp. 123–132. Springer, Boston (2006). https://doi.org/10.1007/978-0-387-37876-3_10
18. Norserium: Plotting graphs in LaTeX—PGFPlots (2015). (in Russian). https://habrahabr.ru/post/250997/
19. Roberts, J.A., Hann, I.H., Slaughter, S.A.: Understanding the motivations, participation, and performance of open source software developers: a longitudinal study of the Apache projects. Manag. Sci. **52**(7), 984–999 (2006). https://doi.org/10.1287/mnsc.1060.0554
20. Schön, D.: IT-Unterstützung, pp. 199–346. Springer Fachmedien Wiesbaden, Wiesbaden (2016). https://doi.org/10.1007/978-3-658-08009-9_5
21. Schwessinger, M., Schürmann, T., Süßer, K.: OLE und Dynamischer Datenaustausch, pp. 793–808. Vieweg+Teubner Verlag, Wiesbaden (1992). https://doi.org/10.1007/978-3-322-96369-7_42
22. Sofroniou, N., Hutcheson, G.D.: Technical typesetting using latex. Behav. Res. Methods Instrum. Comput. **26**(3), 369–371 (1994). https://doi.org/10.3758/BF03204648
23. Sorokin, M.O., Khoteev, S.D.: Free instruments for organization of information systems development in building. Youth Herald Tech. Sci. (1) (2013). (in Russian). http://sntbul.bmstu.ru/doc/532852.html
24. Vukolov, A.: Free and open source software applications for education of TMM discipline in Bauman university, pp. 253–260. Springer, Cham (2017). https://doi.org/10.1007/978-3-319-44156-6_26
25. Wozny, M.J.: Beyond computer graphics and CAD/CAM, pp. 3–9. Springer, Tokyo (1987). https://doi.org/10.1007/978-4-431-68057-4_1

Author Index

J. C. García-Prada and C. Castejón (Eds.): *New Trends in Educational Activity in the Field of Mechanism and Machine Theory*, MMS 64, pp. 217–218, 2019.
https://doi.org/10.1007/978-3-030-00108-7

Printed by Printforce, the Netherlands